Mastering Numerical Computing with NumPy

Master scientific computing and perform complex operations with ease

Umit Mert Cakmak
Mert Cuhadaroglu

BIRMINGHAM - MUMBAI

Mastering Numerical Computing with NumPy

Commissioning Editor: Pravin Dhandre
Acquisition Editor: Viraj Madhav
Content Development Editor: Snehal Kolte
Technical Editor: Dinesh Chaudhary
Copy Editor: Safis Editing
Project Coordinator: Manthan Patel
Proofreader: Safis Editing
Indexer: Priyanka Dhadke
Graphics: Tania Dutta
Production Coordinator: Arvindkumar Gupta

First published: June 2018

Production reference: 1270618

Published by Packt Publishing Ltd.
Livery Place
35 Livery Street
Birmingham
B3 2PB, UK.

ISBN 978-1-78899-335-7

www.packtpub.com

`mapt.io`

Mapt is an online digital library that gives you full access to over 5,000 books and videos, as well as industry leading tools to help you plan your personal development and advance your career. For more information, please visit our website.

Why subscribe?

- Spend less time learning and more time coding with practical eBooks and Videos from over 4,000 industry professionals

- Improve your learning with Skill Plans built especially for you

- Get a free eBook or video every month

- Mapt is fully searchable

- Copy and paste, print, and bookmark content

PacktPub.com

Did you know that Packt offers eBook versions of every book published, with PDF and ePub files available? You can upgrade to the eBook version at `www.PacktPub.com` and as a print book customer, you are entitled to a discount on the eBook copy. Get in touch with us at `service@packtpub.com` for more details.

At `www.PacktPub.com`, you can also read a collection of free technical articles, sign up for a range of free newsletters, and receive exclusive discounts and offers on Packt books and eBooks.

Contributors

About the authors

Umit Mert Cakmak is a data scientist at IBM, where he excels at helping clients solve complex data science problems, from inception to delivery of deployable assets. His research spans multiple disciplines beyond his industry and he likes sharing his insights at conferences, universities, and meet-ups.

> *First and foremost, my heartiest thanks to my mother and father for their true love and support. I am grateful for the lessons that they have taught me. I would like to dedicate my writings to my family, friends, colleagues, and all the great people who relentlessly work to make the world a better place.*

Mert Cuhadaroglu is a BI Developer in EPAM, developing E2E analytics solutions for complex business problems in various industries, mostly investment banking, FMCG, media, communication, and pharma. He consistently uses advanced statistical models and ML algorithms to provide actionable insights. Throughout his career, he has worked in several other industries, such as banking and asset management. He continues his academic research in AI for trading algorithms.

> *I am very grateful to my parents, who have always encouraged me to pursue knowledge. I would also like to thank my coauthor, friends, and the Packt team. I would like to dedicate my writings to my family, friends and all people who supported me in this endeavor.*

About the reviewer

Tiago Antao is a computer scientist turned computational biologist with a PhD from Liverpool School of Tropical Medicine in UK. He is a co-author of Biopython, a major bioinformatics package written in Python.

In his career he has worked at University of Cambridge (UK) and University of Oxford (UK), and is currently a research scientist at University of Montana (USA). He is the author of *Bioinformatics with Python Cookbook*.

Packt is searching for authors like you

If you're interested in becoming an author for Packt, please visit `authors.packtpub.com` and apply today. We have worked with thousands of developers and tech professionals, just like you, to help them share their insight with the global tech community. You can make a general application, apply for a specific hot topic that we are recruiting an author for, or submit your own idea.

Table of Contents

Preface

If you are trying to hone your skills in the field of data science, there are many books and courses out there with varying levels of difficulty. What usually happens is that you start to study introductory resources and then continue with more in-depth, technical ones to get a taste of a new field or technology. If you were following this kind of learning path for sometime, you must have realized that it becomes very time consuming journey. We, as lifelong learners, need books with more compact representation of knowledge and experience which requires the right balance between theory and practice. This book aims to bring beginner, intermediate, and advanced concepts together and it is our humble effort to build up your knowledge from scratch.

This book assumes no previous background of scientific computing and will introduce various subjects using practical examples. It may sometimes feel like separate topics pulled together randomly and the book's flow doesn't stick to one consistent path. This was a deliberate decision we made to give you a little taste of several different topics and applications.

We hope that you will read this book to have a broader overview of scientific computing as well as to master the nitty-gritty of NumPy and other supporting scientific libraries of Python such as SciPy and Scikit-Learn.

Who this book is for

This book is for everyone who would like to gain additional knowledge in the data science field. *Mastering Numerical Computing with NumPy* is for you if you are a Python programmer, data analyst, data engineer, or data science enthusiast who wants to master the intricacies of NumPy and build solutions for your numeric and scientific computational problems. You are expected to have familiarity with mathematics to get the most out of this book.

What this book covers

Chapter 1, *Working with Numpy Arrays*, explains the basics of numerical computing with NumPy, which is a Python library for working with multi-dimensional arrays and matrices used by scientific computing applications.

Chapter 2, *Linear Algebra with Numpy*, covers the basics of linear algebra and provides practical NumPy examples.

Chapter 3, *Exploratory Data Analysis of Boston Housing Data with NumPy Statistics*, explains exploratory data analysis and provides examples using Boston Housing Dataset.

Chapter 4, *Predicting Housing Prices Using Linear Regression*, covers supervised learning and provides a practical example for predicting housing prices using linear regression.

Chapter 5, *Clustering Clients of a Wholesale Distributor Using NumPy*, explains unsupervised learning and provides a practical example of a clustering algorithm to model a wholesale distributor sales dataset, which contains information on annual spending in monetary units for diverse product categories.

Chapter 6, *NumPy, SciPy, Pandas, and Scikit-Learn*, shows the relationship between NumPy and other libraries and provides examples of how they are used together.

Chapter 7, *Advanced Numpy*, explains the advanced considerations of NumPy library usage.

Chapter 8, *Overview of High-Performance Numerical Computing Libraries*, introduces several low-level, high-performance numerical computing libraries and their relationship with NumPy.

Chapter 9, *Performance Benchmarks*, takes a deep dive into the performance of NumPy algorithms depending on the underlying high-performance numerical computing libraries.

To get the most out of this book

1. Basic Python programming knowledge will definitely help, though it is not strictly necessary
2. Anaconda distribution for Python 3 will be enough to cover most of the examples used in this book

Download the example code files

You can download the example code files for this book from your account at www.packtpub.com. If you purchased this book elsewhere, you can visit www.packtpub.com/support and register to have the files emailed directly to you.

You can download the code files by following these steps:

1. Log in or register at `www.packtpub.com`.
2. Select the **SUPPORT** tab.
3. Click on **Code Downloads & Errata**.
4. Enter the name of the book in the **Search** box and follow the onscreen instructions.

Once the file is downloaded, please make sure that you unzip or extract the folder using the latest version of:

- WinRAR/7-Zip for Windows
- Zipeg/iZip/UnRarX for Mac
- 7-Zip/PeaZip for Linux

The code bundle for the book is also hosted on GitHub at `https://github.com/PacktPublishing/Mastering-Numerical-Computing-with-NumPy`. In case there's an update to the code, it will be updated on the existing GitHub repository.

We also have other code bundles from our rich catalog of books and videos available at `https://github.com/PacktPublishing/`. Check them out!

Download the color images

We also provide a PDF file that has color images of the screenshots/diagrams used in this book. You can download it here: `http://www.packtpub.com/sites/default/files/downloads/MasteringNumericalComputingwithNumPy_ColorImages.pdf`.

Conventions used

There are a number of text conventions used throughout this book.

`CodeInText`: Indicates code words in text, database table names, folder names, filenames, file extensions, pathnames, dummy URLs, user input, and Twitter handles. Here is an example: "Another important parameter in this function is `learning_rate`."

A block of code is set as follows:

```
'sepal width (cm)',
'petal length (cm)',
'petal width (cm)'])
```

Any command-line input or output is written as follows:

```
$ sudo apt-get update
$ sudo apt-get upgrade
```

Bold: Indicates a new term, an important word, or words that you see onscreen. For example, words in menus or dialog boxes appear in the text like this. Here is an example: "**Dependent** is the variable that we want to predict."

 Warnings or important notes appear like this.

 Tips and tricks appear like this.

Get in touch

Feedback from our readers is always welcome.

General feedback: Email `feedback@packtpub.com` and mention the book title in the subject of your message. If you have questions about any aspect of this book, please email us at `questions@packtpub.com`.

Errata: Although we have taken every care to ensure the accuracy of our content, mistakes do happen. If you have found a mistake in this book, we would be grateful if you would report this to us. Please visit `www.packtpub.com/submit-errata`, selecting your book, clicking on the Errata Submission Form link, and entering the details.

Piracy: If you come across any illegal copies of our works in any form on the Internet, we would be grateful if you would provide us with the location address or website name. Please contact us at copyright@packtpub.com with a link to the material.

If you are interested in becoming an author: If there is a topic that you have expertise in and you are interested in either writing or contributing to a book, please visit authors.packtpub.com.

Reviews

Please leave a review. Once you have read and used this book, why not leave a review on the site that you purchased it from? Potential readers can then see and use your unbiased opinion to make purchase decisions, we at Packt can understand what you think about our products, and our authors can see your feedback on their book. Thank you!

For more information about Packt, please visit packtpub.com.

Working with NumPy Arrays 1

Scientific computing is a multidisciplinary field, with its applications spanning across disciplines such as numerical analysis, computational finance, and bioinformatics.

Let's consider a case for financial markets. When you think about financial markets, there is a huge interconnected web of interactions. Governments, banks, investment funds, insurance companies, pensions, individual investors, and others are involved in this exchange of financial instruments. You can't simply model all the interactions between market participants because everyone who is involved in financial transactions has different motives and different risk/return objectives. There are also other factors which affect the prices of financial assets. Even modeling one asset price requires you to do a tremendous amount of work, and your success is not guaranteed. In mathematical terms, this doesn't have a closed-form solution and this makes a great case for utilizing scientific computing where you can use advanced computing techniques to attack such problems.

By writing computer programs, you will have the power to better understand the system you are working on. Usually, the computer program you will be writing will be some sort of simulation, such as the Monte Carlo simulation. By using a simulation such as Monte Carlo, you can model the price of option contracts. Pricing financial assets is a good material for simulations, simply because of the complexity of financial markets. All of these mathematical computations need a powerful, scalable and convenient structure for your data (which is mostly in matrix form) when you do your computation. In other words, you need a more compact structure than a *list* in order to simplify your task. NumPy is a perfect candidate for performant vector/matrix operations and its extensive library of mathematical operations makes numeric computing easy and efficient.

In this chapter, we will cover the following topics:

- The importance of NumPy
- Theoretical and practical information about vectors and matrices
- NumPy array operations and their usage in multidimensional arrays

The question is, where should we start practicing coding skills? In this book, you will be using Python because of its huge adoption in the scientific community, and you will mainly work with a specific library called NumPy, which stands for numerical Python.

Technical requirements

In this book, we will use Jupyter Notebooks. We will edit and run Python code via a web browser. It's an open source platform which you can install by following the instructions in this link: `http://jupyter.org/install`.

This book will be using Python 3.x, so when you open a new notebook, you should pick Python 3 kernel. Alternatively, you can install Jupyter Notebook using Anaconda (Python version 3.6), which is highly recommended. You can install it by following the instructions in this link: `https://www.anaconda.com/download/`.

Why do we need NumPy?

Python has become a rockstar programming language recently, not only because it has friendly syntax and readability, but because it can be used for a variety of purposes. Python's ecosystem of various libraries makes various computations relatively easy for programmers. Stack Overflow is one the most popular websites for programmers. Users can ask questions by tagging which programming language they relate to. The following figure shows the growth of major programming languages by calculating these tags and plot the popularity of major programming languages over the years. The research conducted by Stack Overflow can be further analyzed via this link to their official blog: `https://stackoverflow.blog/2017/09/06/incredible-growth-python/`:

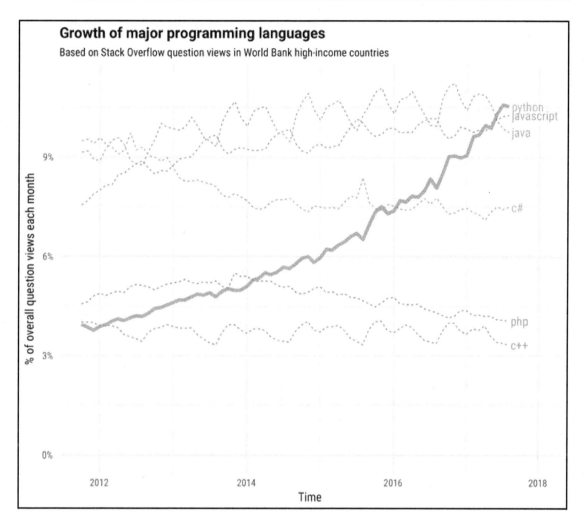

Growth of major programming languages

NumPy is the most fundamental package for scientific computing in Python and is the base for many other packages. Since Python was not initially designed for numerical computing, this need has arised in the late 90's when Python started to become popular among engineers and programmers who needed faster vector operations. As you can see from the following figure, many popular machine learning and computational packages use some of NumPy's features, and the most important thing is that they use NumPy arrays heavily in their methods, which makes NumPy an essential library for scientific projects.

The figure shows some well-known libraries which use NumPy features:

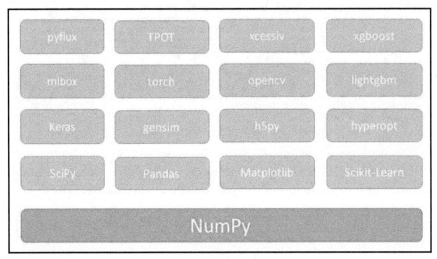

NumPy stack

For numerical computing, you mainly work with vectors and matrices. You can manipulate them in different ways by using a range of mathematical functions. NumPy is a perfect fit for these kinds of situations since it allows users to have their computations completed efficiently. Even though Python lists are very easy to create and manipulate, they don't support *vectorized* operations. Python doesn't have fixed type elements in lists and for example, for loop is not very efficient because, at every iteration, data type needs to be checked. In NumPy arrays, however, the data type is fixed and also supports *vectorized* operations. NumPy is not just more efficient in multidimensional array operations comparing to Python lists; it also provides many mathematical methods that you can apply as soon as it's imported. NumPy is a core library for the scientific Python data science stack.

SciPy has strong relationship with NumPy as it's using NumPy multidimensional arrays as a base data structure for its scientific functions for linear algebra, optimization. interpolation, integration, FFT, signal and image processing and others. SciPy was built on top of the NumPy array framework and uplifted scientific programming with its advanced mathematical functions. Therefore some parts of the NumPy API have been moved to SciPy. This relationship with NumPy makes SciPy more convenient for advanced scientific computing in many cases.

To sum this up, we can summarize NumPy's advantages as follows:

- It's open source and zero-cost
- It's a high-level programming language with user-friendly syntax
- It's more efficient than Python lists
- It has more advanced built-in functions and is well-integrated with other libraries

Who uses NumPy?

In both academic and business circles, you will hear people talking about the tools and technologies they use in their work. Depending on the environment and conditions, you might need to work with specific technologies. For example, if your company has already invested in SAS, you will need to carry out your project in the SAS development environment suited to your problem.

However, one of the advantages of NumPy is that it's open source, and it costs nothing for you to utilize it in your project. If you have already coded in Python, it's super easy to learn. If performance is your concern, you can easily embed C or Fortran code. Moreover, it will introduce you to a whole other set of libraries such as SciPy and Scikit-learn, which you can use to solve almost any problem.

Since data mining and predictive analytics became really important recently, roles like *Data Scientist* and *Data Analyst* are mentioned as the hottest jobs of the 21st century in many business journals such as Forbes, Bloomberg, and so on. People who need to work with data and do analysis, modeling, or forecasting should become familiar with NumPy's usage and its capabilities, as it will help you quickly prototype and test your ideas. If you are a working professional, your firm most probably wants to use data analysis methods in order to move one step ahead of its competitors. If they can better understand the data they have, they can understand the business better, and this will lead them to make better decisions. NumPy plays a critical role here as it is capable of performing wide range of operations and making your projects timewise efficient.

Introduction to vectors and matrices

A matrix is a group of numbers or elements which are arranged as a rectangular array. The matrix's rows and columns are usually indexed by letter. For a *n x m* matrix, n represents the number of rows and m represents the number of columns. If we have a hypothetical *n x m* matrix, it will be structured as follows:

$$X = \begin{bmatrix} x_{11} & . & . & . & x_{1m} \\ . & . & . & . & . \\ . & . & . & . & . \\ . & . & . & . & . \\ x_{n1} & . & . & . & x_{nm} \end{bmatrix}$$

If *n = m*, then it is called a square matrix:

$$X = \begin{bmatrix} x_{11} & x_{12} \\ x_{21} & x_{22} \end{bmatrix}$$

A vector is actually a matrix with one row or one column with more than one element. It can also be defined as the *1-by-m* or *n-by-1* matrix. You can interpret a vector as an arrow or direction in an *m* dimensional space. Generally, the capital letter denotes a matrix, like *X* in this example, and lowercase letters with subscripts like X_{11} denote the element of the matrix *X*.

In addition, there are some important special matrices: the zero matrix (null matrix) and the identity matrix. *0* denotes the zero matrix, which is a matrix of all *0*s (MacDufee 1943 p.27). In a *0* matrix, it's optional to add subscripts:

$$0_{2,2} = \begin{bmatrix} 0 & 0 \\ 0 & 0 \end{bmatrix}$$

The identity matrix denoted by *I*, and its diagonal elements are *1* while the others are *0*:

$$I_{3,3} = \begin{bmatrix} 1 & 0 & 0 \\ 0 & 1 & 0 \\ 0 & 0 & 1 \end{bmatrix}$$

When you multiply a matrix X with the identity matrix, the result will be equal to X:

$$X \times I = X$$

An identity matrix is very useful for calculating the inverse of a matrix. When you multiply any given matrix with its inverse, the result will be an identity matrix:

$$X^{-1} \times X = I$$

Let's briefly see the matrix algebra on NumPy arrays. Addition and subtraction operations for matrices are similar to math equations with ordinary single numbers. As an example:

$$\begin{bmatrix} 1 & 4 & 7 \\ 2 & 5 & 8 \end{bmatrix} + \begin{bmatrix} 10 & 14 & 16 \\ 13 & 18 & 21 \end{bmatrix} = \begin{bmatrix} 1+10 & 4+14 & 7+16 \\ 2+13 & 5+18 & 8+21 \end{bmatrix}$$

$$= \begin{bmatrix} 11 & 18 & 23 \\ 15 & 23 & 29 \end{bmatrix}$$

Scalar multiplication is also pretty straightforward. As an example, if you multiply your matrix X by 4, the only thing that you should do is multiply each element with the value 4 as follows:

$$X = \begin{bmatrix} 1 & 2 \\ 3 & 4 \end{bmatrix}$$

$$4X = 4. \begin{bmatrix} 1 & 2 \\ 3 & 4 \end{bmatrix} = \begin{bmatrix} 4 & 8 \\ 12 & 16 \end{bmatrix}$$

The seemingly complicated part of matrix manipulation at the beginning is matrix multiplication.

Imagine you have two matrices as X and Y, where X is an $a \times b$ matrix and Y is $b \times c$ an matrix:

$$X = \begin{bmatrix} X_{11} & X_{12} & X_{13} & . & . & X_{1b} \\ X_{21} & X_{22} & X_{23} & . & . & X_{2b} \\ X_{31} & X_{32} & X_{33} & . & . & X_{3b} \\ . & . & . & . & . & . \\ . & . & . & . & . & . \\ X_{a1} & X_{a2} & X_{a3} & . & . & X_{ab} \end{bmatrix}$$

$$Y = \begin{bmatrix} Y_{11} & Y_{12} & Y_{13} & . & . & Y_{1c} \\ Y_{21} & Y_{22} & Y_{23} & . & . & Y_{2c} \\ Y_{31} & Y_{32} & Y_{33} & . & . & Y_{3c} \\ . & . & . & . & . & . \\ . & . & . & . & . & . \\ Y_{b1} & Y_{b2} & Y_{b3} & . & . & Y_{bc} \end{bmatrix}$$

The product of these two matrices will be as follows:

$$Z = \begin{bmatrix} z_{11} & z_{12} & z_{13} & . & . & z_{1c} \\ z_{21} & z_{22} & z_{23} & . & . & z_{2c} \\ z_{31} & z_{32} & z_{33} & . & . & z_{3c} \\ . & . & . & . & . & . \\ . & . & . & . & . & . \\ z_{a1} & z_{a2} & z_{a3} & . & . & z_{ac} \end{bmatrix}$$

So each element of the product matrix is calculated as follows:

$$Z_{ij} = x_{i1}y_{1j} + \ldots + x_{ib}y_{bj} = \sum_{k=1}^{b} x_{ib}y_{bj}$$

Don't worry if you didn't understand the notation. The following example will make things clearer. You have matrices X and Y and the goal is to get the matrix product of these matrices:

$$X = \begin{bmatrix} 1 & 0 & 4 \\ 3 & 3 & 1 \end{bmatrix}, Y = \begin{bmatrix} 2 & 5 \\ 1 & 1 \\ 3 & 2 \end{bmatrix}$$

The basic idea is that the product of the i_{th} row of X and the j_{th} of column Y will become the i_{th}, j_{th} element of the matrix in the result. Multiplication will start with the first row of X and the first column of Y, so their product will be Z[1,1]:

$$Z = \begin{bmatrix} (1 \times 2) + (0 \times 1) + (4 \times 3) & (1 \times 5) + (0 \times 1) + (4 \times 2) \\ (3 \times 2) + (3 \times 1) + (1 \times 3) & (3 \times 5) + (3 \times 1) + (1 \times 2) \end{bmatrix} = \begin{bmatrix} 14 & 13 \\ 12 & 20 \end{bmatrix}$$

You can cross-check the results easily with the following four lines of code:

```
In [1]: import numpy as np
        x = np.array([[1,0,4],[3,3,1]])
        y = np.array([[2,5],[1,1],[3,2]])
        x.dot(y)
Out[1]: array([[14, 13],[12, 20]])
```

The previous code block is just a demonstration of how easy to calculate the dot product of two matrices by use of NumPy. In later chapters, we will go more in deep into matrix operations and linear algebra.

Basics of NumPy array objects

As mentioned in the preceding section, what makes NumPy special is the usage of multidimensional arrays called **ndarrays**. All ndarray items are homogeneous and use the same size in memory. Let's start by importing NumPy and analyzing the structure of a NumPy array object by creating the array. You can easily import this library by typing the following statement into your console. You can use any naming convention instead of np, but in this book, np will be used as it's the standard convention. Let's create a simple array and explain what the attributes hold by Python behind the scenes as metadata of the created array, so-called **attributes:**

```
In [2]: import numpy as np
        x = np.array([[1,2,3],[4,5,6]])
        x
Out[2]: array([[1, 2, 3],[4, 5, 6]])
In [3]: print("We just create a ", type(x))
Out[3]: We just create a <class 'numpy.ndarray'>
In [4]: print("Our template has shape as" ,x.shape)
Out[4]: Our template has shape as (2, 3)
In [5]: print("Total size is",x.size)
Out[5]: Total size is 6
In [6]: print("The dimension of our array is " ,x.ndim)
Out[6]: The dimension of our array is 2
In [7]: print("Data type of elements are",x.dtype)
Out[7]: Data type of elements are int32
In [8]: print("It consumes",x.nbytes,"bytes")
Out[8]: It consumes 24 bytes
```

As you can see, the type of our object is a NumPy array. `x.shape` returns a tuple which gives you the dimension of the array as an output such as *(n,m)*. You can get the total number of elements in an array by using `x.size`. In our example, we have six elements in total. Knowing attributes such as *shape and dimension* is very important. The more you know, the more you will be comfortable with computations. If you don't know your array's size and dimensions, it wouldn't be wise to start doing computations with it. In NumPy, you can use `x.ndim` to check what the dimension of your array is. There are other attributes such as `dtype` and `nbytes`, which are very useful while you are checking memory consumption and verifying what kind of data type you should use in the array. In our example, each element has a data type of `int32` that consumes 24 bytes in total. You can also force some of these attributes while creating your array such as `dtype`. Previously, the data type was an integer. Let's switch it to `float`, `complex`, or `uint` (unsigned integer). In order to see what the data type change does, let's analyze what byte consumption is, which is shown as follows:

```
In [9]: x = np.array([[1,2,3],[4,5,6]], dtype = np.float)
        print(x)
Out[9]: print(x.nbytes)
        [[ 1. 2. 3.]
         [ 4. 5. 6.]]
        48
In [10]: x = np.array([[1,2,3],[4,5,6]], dtype = np.complex)
         print(x)
         print(x.nbytes)
Out[10]: [[ 1.+0.j 2.+0.j 3.+0.j]
          [ 4.+0.j 5.+0.j 6.+0.j]]
         96
In [11]: x = np.array([[1,2,3],[4,-5,6]], dtype = np.uint32)
         print(x)
         print(x.nbytes)
Out[11]: [[ 1 2 3]
          [ 4 4294967291 6]]
         24
```

As you can see, each type consumes a different number of bytes. Imagine you have a matrix as follows and that you are using `int64` or `int32` as the data type:

```
In [12]: x = np.array([[1,2,3],[4,5,6]], dtype = np.int64)
         print("int64 consumes",x.nbytes, "bytes")
         x = np.array([[1,2,3],[4,5,6]], dtype = np.int32)
         print("int32 consumes",x.nbytes, "bytes")
Out[12]: int64 consumes 48 bytes
         int32 consumes 24 bytes
```

The memory need is doubled if you use `int64`. Ask this question to yourself; which data type would suffice? Until your numbers are higher than 2,147,483,648 and lower than -2,147,483,647, using `int32` is enough. Imagine you have a huge array with a size over 100 MB. In such cases, this conversion plays a crucial role in performance.

As you may have noticed in the previous example, when you were changing the data types, you were creating an array each time. Technically, you cannot change the `dtype` after you create the array. However, what you can do is either create it again or copy the existing one with a new `dtype` and with the `astype` attribute. Let's create a copy of the array with the new `dtype`. Here is an example of how you can also change your `dtype` with the `astype` attribute:

```
In [13]: x_copy = np.array(x, dtype = np.float)
         x_copy
Out[13]: array([[ 1., 2., 3.],
         [ 4., 5., 6.]])
In [14]: x_copy_int = x_copy.astype(np.int)
         x_copy_int
Out[14]: array([[1, 2, 3],
         [4, 5, 6]])
```

Please keep in mind that when you use the `astype` attribute, it doesn't change the `dtype` of the `x_copy`, even though you applied it to `x_copy`. It keeps the `x_copy`, but creates a `x_copy_int`:

```
In [15]: x_copy
Out[15]: array([[ 1., 2., 3.],
         [ 4., 5., 6.]])
```

Let's imagine a case where you are working in a research group that tries to identify and calculate the risks of an individual patient who has cancer. You have 100,000 records (rows), where each row represents a single patient, and each patient has 100 features (results of some of the tests). As a result, you have (100000, 100) arrays:

```
In [16]: Data_Cancer= np.random.rand(100000,100)
         print(type(Data_Cancer))
         print(Data_Cancer.dtype)
         print(Data_Cancer.nbytes)
         Data_Cancer_New = np.array(Data_Cancer, dtype = np.float32)
         print(Data_Cancer_New.nbytes)
Out[16]: <class 'numpy.ndarray'>
         float64
         80000000
         40000000
```

As you can see from the preceding code, their size decreases from 80 MB to 40 MB just by changing the `dtype`. What we get in return is less precision after decimal points. Instead of being precise to 16 decimals points, you will have only 7 decimals. In some machine learning algorithms, precision can be negligible. In such cases, you should feel free to adjust your `dtype` so that it minimizes your memory usage.

NumPy array operations

This section will guide you through the creation and manipulation of numerical data with NumPy. Let's start by creating a NumPy array from the list:

```
In [17]: my_list = [2, 14, 6, 8]
         my_array = np.asarray(my_list)
         type(my_array)
Out[17]: numpy.ndarray
```

Let's do some addition, subtraction, multiplication, and division with scalar values:

```
In [18]: my_array + 2
Out[18]: array([ 4, 16, 8, 10])
In [19]: my_array - 1
Out[19]: array([ 1, 13, 5, 7])
In [20]: my_array * 2
Out[20]: array([ 4, 28, 12, 16, 8])
In [21]: my_array / 2
Out[21]: array([ 1. , 7. , 3. , 4. ])
```

It's much harder to do the same operations in a list because the list does not support vectorized operations and you need to iterate its elements. There are many ways to create NumPy arrays, and now you will use one of these methods to create an array which is full of zeros. Later, you will perform some arithmetic operations to see how NumPy behaves in element-wise operations between two arrays:

```
In [22]: second_array = np.zeros(4) + 3
         second_array
Out[22]: array([ 3., 3., 3., 3.])
In [23]: my_array - second_array
Out[23]: array([ -1., 11., 3., 5.])
In [24]: second_array / my_array
Out[24]: array([ 1.5 , 0.21428571, 0.5 , 0.375 ])
```

As we did in the previous code, you can create an array which is full of ones with `np.ones` or an identity array with `np.identity` and do the same algebraic operations that you did previously:

```
In [25]: second_array = np.ones(4) + 3
         second_array
Out[25]: array([ 4.,  4.,  4.,  4.])
In [26]: my_array - second_array
Out[26]: array([ -2., 10.,  2.,  4.])
In [27]: second_array / my_array
Out[27]: array([ 2. , 0.28571429, 0.66666667, 0.5 ])
```

It works as expected with the `np.ones` method, but when you use the identity matrix, the calculation returns a *(4,4)* array as follows:

```
In [28]: second_array = np.identity(4)
         second_array
Out[28]: array([[ 1.,  0.,  0.,  0.],
                [ 0.,  1.,  0.,  0.],
                [ 0.,  0.,  1.,  0.],
                [ 0.,  0.,  0.,  1.]])
In [29]: second_array = np.identity(4) + 3
         second_array
Out[29]: array([[ 4.,  3.,  3.,  3.],
                [ 3.,  4.,  3.,  3.],
                [ 3.,  3.,  4.,  3.],
                [ 3.,  3.,  3.,  4.]])
In [30]: my_array - second_array
Out[30]: array([[ -2., 11.,  3.,  5.],
                [ -1., 10.,  3.,  5.],
                [ -1., 11.,  2.,  5.],
                [ -1., 11.,  3.,  4.]])
```

What this does is subtract the first element of `my_array` from all of the elements of the first column of `second_array` and the `second_element` of the second column, and so on. The same rule is applied to division as well. Please keep in mind that you can successfully do array operations even if they are not exactly the same shape. Later in this chapter, you will learn about broadcasting errors when computation cannot be done between two arrays due to differences in their shapes:

```
In [31]: second_array / my_array
Out[31]: array([[ 2.  , 0.21428571, 0.5       , 0.375     ],
                [ 1.5 , 0.28571429, 0.5       , 0.375     ],
                [ 1.5 , 0.21428571, 0.66666667, 0.375     ],
                [ 1.5 , 0.21428571, 0.5       , 0.5       ]])
```

One of the most useful methods in creating NumPy arrays is `arange`. This returns an array for a given interval between your start and end values. The first argument is the start value of your array, the second is the end value (where it stops creating values), and the third one is the interval. Optionally, you can define your `dtype` as the fourth argument. The default interval values are 1:

```
In [32]: x = np.arange(3,7,0.5)
         x
Out[32]: array([ 3. , 3.5, 4. , 4.5, 5. , 5.5, 6. , 6.5])
```

There is another way to create an array with fixed intervals between the start and stop point when you cannot decide what the interval should be, but you should know how many splits your array should have:

```
In [33]: x = np.linspace(1.2, 40.5, num=20)
         x
Out[33]: array([ 1.2        , 3.26842105, 5.33684211, 7.40526316,
9.47368421,
                11.54210526, 13.61052632, 15.67894737, 17.74736842,
19.81578947,
                21.88421053, 23.95263158, 26.02105263, 28.08947368,
30.15789474,
                32.22631579, 34.29473684, 36.36315789, 38.43157895, 40.5
])
```

There are two different methods which are similar in usage but return different sequences of numbers because their base scale is different. This means that the distribution of the numbers will be different as well. The first one is `geomspace`, which returns numbers on a logarithmic scale with a geometric progression:

```
In [34]: np.geomspace(1, 625, num=5)
Out[34]: array([ 1., 5., 25., 125., 625.])
```

The other important method is `logspace`, where you can return the values for your start and stop points, which are evenly scaled in:

```
In [35]: np.logspace(3, 4, num=5)
Out[35]: array([ 1000. , 1778.27941004, 3162.27766017, 5623.4132519 ,
                10000. ])
```

What are these arguments? If the starting point is 3 and the ending point is 4, then these functions return the numbers which are much higher than the initial range. Your starting point is actually set as default to `10**Start Argument and the ending is set as 10**End Argument`. So technically, in this example, the starting point is `10**3` and the ending point is `10**4`. You can avoid such situations and keep your start and end points the same as when you put them as arguments in the method. The trick is to use base 10 logarithms of the given arguments:

```
In [36]: np.logspace(np.log10(3) , np.log10(4) , num=5)
Out[36]: array([ 3. , 3.2237098 , 3.46410162, 3.72241944, 4. ])
```

By now, you should be familiar with different ways of creating arrays with different distributions. You have also learned how to do some basic operations with them. Let's continue with other useful functions that you will definitely use in your day to day work. Most of the time, you will have to work with multiple arrays and you will need to compare them very quickly. NumPy has a great solution for this problem; you can compare the arrays as you would compare two integers:

```
In [37]: x = np.array([1,2,3,4])
         y = np.array([1,3,4,4])
         x == y
Out[37]: array([ True, False, False, True], dtype=bool)
```

The comparison is done element-wise and it returns a Boolean vector, whether elements are matching in two different arrays or not. This method works well in small size arrays and also gives you more details. You can see from the output array where the values are represented as False, that these indexed values are not matching in these two arrays. If you have a large array, you may also choose to get a single answer to your question, whether the elements are matching in two different arrays or not:

```
In [38]: x = np.array([1,2,3,4])
         y = np.array([1,3,4,4])
         np.array_equal(x,y)
Out[38]: False
```

Here, you have a single Boolean output. You only know that arrays are not equal, but you don't know which elements exactly are not equal. The comparison is not only limited to checking whether two arrays are equal or not. You can also do element-wise higher- lower comparison between two arrays:

```
In [39]: x = np.array([1,2,3,4])
         y = np.array([1,3,4,4])
         x < y
Out[39]: array([False, True, True, False], dtype=bool)
```

When you need to do logical comparison (AND, OR, XOR), you can use them in your array as follows:

```
In [40]: x = np.array([0, 1, 0, 0], dtype=bool)
         y = np.array([1, 1, 0, 1], dtype=bool)
         np.logical_or(x,y)
Out[40]: array([ True, True, False, True], dtype=bool)
In [41]: np.logical_and(x,y)
Out[41]: array([False, True, False, False], dtype=bool)
In [42]: x = np.array([12,16,57,11])
         np.logical_or(x < 13, x > 50)
Out[42]: array([ True, False, True, True], dtype=bool)
```

So far, algebraic operations such as addition and multiplication have been covered. How can we use these operations with transcendental functions such as the exponential function, logarithms, or trigonometric functions?

```
In [43]: x = np.array([1, 2, 3,4 ])
         np.exp(x)
Out[43]: array([ 2.71828183, 7.3890561 , 20.08553692, 54.59815003])
In [44]: np.log(x)
Out[44]: array([ 0. , 0.69314718, 1.09861229, 1.38629436])
In [45]: np.sin(x)
Out[45]: array([ 0.84147098, 0.90929743, 0.14112001, -0.7568025 ])
```

What about the transpose of a matrix? First, you will use the reshape function with arange to set the desired shape of the matrix:

```
In [46]: x = np.arange(9)
         x
Out[46]: array([0, 1, 2, 3, 4, 5, 6, 7, 8])
In [47]: x = np.arange(9).reshape((3, 3))
         x
Out[47]: array([[0, 1, 2],
                [3, 4, 5],
                [6, 7, 8]])
In [48]: x.T
Out[48]: array([[0, 3, 6],
                [1, 4, 7],
                [2, 5, 8]])
```

You transpose the 3*3 array, so the shape doesn't change because both dimensions are 3. Let's see what happens when you don't have a square array:

```
In [49]: x = np.arange(6).reshape(2,3)
         x
Out[49]: array([[0, 1, 2],
```

```
                    [3, 4, 5]])
In [50]: x.T
Out[50]: array([[0, 3],
                [1, 4],
                [2, 5]])
```

The transpose works as expected and the dimensions are switched as well. You can also get summary statistics from arrays such as mean, median, and standard deviation. Let's start with methods that NumPy offers for calculating basic statistics:

Method	Description
np.sum	Returns the sum of all array values or along the specified axis
np.amin	Returns the minimum value of all arrays or along the specified axis
np.amax	Returns the maximum value of all arrays or along the specified axis
np.percentile	Returns the given q^{th} percentile of all arrays or along the specified axis
np.nanmin	The same as np.amin, but ignores NaN values in an array
np.nanmax	The same as np.amax, but ignores NaN values in an array
np.nanpercentile	The same as np.percentile, but ignores NaN values in an array

The following code block gives an example of the preceding statistical methods of NumPy. These methods are very useful as you can operate the methods in a whole array or axis-wise according to your needs. You should note that you can find more fully-featured and better implementations of these methods in SciPy as it uses NumPy multidimensional arrays as a data structure:

```
In [51]: x = np.arange(9).reshape((3,3))
         x
Out[51]: array([[0, 1, 2],
                [3, 4, 5],
                [6, 7, 8]])
In [52]: np.sum(x)
Out[52]: 36
In [53]: np.amin(x)
Out[53]: 0
In [54]: np.amax(x)
Out[54]: 8
In [55]: np.amin(x, axis=0)
Out[55]: array([0, 1, 2])
In [56]: np.amin(x, axis=1)
Out[56]: array([0, 3, 6])
In [57]: np.percentile(x, 80)
Out[57]: 6.400000000000004
```

The axis argument determines the dimension that this function will operate on. In this example, axis=0 represents the first axis which is the *x* axis, and axis = 1 represents the second axis which is y. When we use a regular amin(x), we return a single value because it calculates the minimum value in all arrays, but when we specify the axis, it starts evaluating the function axis-wise and returns an array which shows the results for each row or column. Imagine you have a large array; you find the max value by using amax, but what will happen if you need to pass the index of this value to another function? In such cases, argmin and argmax come to the rescue, as shown in the following snippet:

```
In [58]: x = np.array([1,-21,3,-3])
         np.argmax(x)
Out[58]: 2
In [59]: np.argmin(x)
Out[59]: 1
```

Let's continue with more statistical functions:

Method	Description
np.mean	Returns the mean of all array values or along the specific axis
np.median	Returns the median of all array values or along the specific axis
np.std	Returns the standard deviation of all array values or along the specific axis
np.nanmean	The same as np.mean, but ignores NaN values in an array
np.nanmedian	The same as np.nanmedian, but ignores NaN values in an array
np.nonstd	The same as np.nanstd, but ignores NaN values in an array

The following code gives more examples of the preceding statistical methods of NumPy. These methods are heavily used in data discovery phases, where you analyze your data features and distribution:

```
In [60]: x = np.array([[2, 3, 5], [20, 12, 4]])
         x
Out[60]: array([[ 2, 3, 5],
                [20, 12, 4]])
In [61]: np.mean(x)
Out[61]: 7.666666666666667
In [62]: np.mean(x, axis=0)
Out[62]: array([ 11. , 7.5, 4.5])
In [63]: np.mean(x, axis=1)
Out[63]: array([ 3.33333333, 12. ])
In [64]: np.median(x)
Out[64]: 4.5
In [65]: np.std(x)
Out[65]: 6.3944420310836261
```

Working with multidimensional arrays

This section will give you a brief understanding of multidimensional arrays by going through different matrix operations.

In order to do matrix multiplication in NumPy, you have to use dot () instead of *. Let's see some examples:

```
In [66]: c = np.ones((4, 4))
         c*c
Out[66]: array([[ 1., 1., 1., 1.],
                [ 1., 1., 1., 1.],
                [ 1., 1., 1., 1.],
                [ 1., 1., 1., 1.]])
In [67]: c.dot(c)
Out[67]: array([[ 4., 4., 4., 4.],
                [ 4., 4., 4., 4.],
                [ 4., 4., 4., 4.],
                [ 4., 4., 4., 4.]])
```

The most important topic in working with multidimensional arrays is stacking, in other words how to merge two arrays. hstack is used for stacking arrays horizontally (column-wise) and vstack is used for stacking arrays vertically (row-wise). You can also split the columns with the hsplit and vsplit methods in the same way that you stacked them:

```
In [68]: y = np.arange(15).reshape(3,5)
         x = np.arange(10).reshape(2,5)
         new_array = np.vstack((y,x))
         new_array
Out[68]: array([[ 0, 1, 2, 3, 4],
                [ 5, 6, 7, 8, 9],
                [10, 11, 12, 13, 14],
                [ 0, 1, 2, 3, 4],
                [ 5, 6, 7, 8, 9]])
In [69]: y = np.arange(15).reshape(5,3)
         x = np.arange(10).reshape(5,2)
         new_array = np.hstack((y,x))
         new_array
Out[69]: array([[ 0, 1, 2, 0, 1],
                [ 3, 4, 5, 2, 3],
                [ 6, 7, 8, 4, 5],
                [ 9, 10, 11, 6, 7],
                [12, 13, 14, 8, 9]])
```

These methods are very useful in machine learning applications, especially when creating datasets. After you stack your arrays, you can check their descriptive statistics by using `Scipy.stats`. Imagine a case where you have 100 records, and each record has 10 features, which means you have a 2D matrix which has 100 rows and 10 columns. The following example shows how you can easily get some descriptive statistics for each feature:

```
In [70]: from scipy import stats
         x= np.random.rand(100,10)
         n, min_max, mean, var, skew, kurt = stats.describe(x)
         new_array = np.vstack((mean,var,skew,kurt,min_max[0],min_max[1]))
         new_array.T
Out[70]: array([[ 5.46011575e-01,  8.30007104e-02, -9.72899085e-02,
                  -1.17492785e+00,  4.07031246e-04,  9.85652100e-01],
                [ 4.79292653e-01,  8.13883169e-02,  1.00411352e-01,
                  -1.15988275e+00,  1.27241020e-02,  9.85985488e-01],
                [ 4.81319367e-01,  8.34107619e-02,  5.55926602e-02,
                  -1.20006450e+00,  7.49534810e-03,  9.86671083e-01],
                [ 5.26977277e-01,  9.33829059e-02, -1.12640661e-01,
                  -1.19955646e+00,  5.74237697e-03,  9.94980830e-01],
                [ 5.42622228e-01,  8.92615897e-02, -1.79102183e-01,
                  -1.13744108e+00,  2.27821933e-03,  9.93861532e-01],
                [ 4.84397369e-01,  9.18274523e-02,  2.33663872e-01,
                  -1.36827574e+00,  1.18986562e-02,  9.96563489e-01],
                [ 4.41436165e-01,  9.54357485e-02,  3.48194314e-01,
                  -1.15588500e+00,  1.77608372e-03,  9.93865324e-01],
                [ 5.34834409e-01,  7.61735119e-02, -2.10467450e-01,
                  -1.01442389e+00,  2.44706226e-02,  9.97784091e-01],
                [ 4.90262346e-01,  9.28757119e-02,  1.02682367e-01,
                  -1.28987137e+00,  2.97705706e-03,  9.98205307e-01],
                [ 4.42767478e-01,  7.32159267e-02,  1.74375646e-01,
                  -9.58660574e-01,  5.52410464e-04,  9.95383732e-01]])
```

NumPy has a great module named `numpy.ma`, which is used for masking array elements. It's very useful when you want to mask (ignore) some elements while doing your calculations. When NumPy masks, it will be treated as an invalid and does not take into account computation:

```
In [71]: import numpy.ma as ma
         x = np.arange(6)
         print(x.mean())
         masked_array = ma.masked_array(x, mask=[1,0,0,0,0,0])
         masked_array.mean()
         2.5
Out[71]: 3.0
```

In the preceding code, you have an array x = [0,1,2,3,4,5]. What you do is mask the first element of the array and then calculate the mean. When an element is masked as 1(True), the associated index value in the array will be masked. This method is also very useful while replacing the NAN values:

```
In [72]: x = np.arange(25, dtype = float).reshape(5,5)
         x[x<5] = np.nan
         x
Out[72]: array([[ nan,  nan,  nan,  nan,  nan],
                [  5.,   6.,   7.,   8.,   9.],
                [ 10.,  11.,  12.,  13.,  14.],
                [ 15.,  16.,  17.,  18.,  19.],
                [ 20.,  21.,  22.,  23.,  24.]])
In [73]: np.where(np.isnan(x), ma.array(x, mask=np.isnan(x)).mean(axis=0),
         x)
Out[73]: array([[ 12.5,  13.5,  14.5,  15.5,  16.5],
                [  5. ,   6. ,   7. ,   8. ,   9. ],
                [ 10. ,  11. ,  12. ,  13. ,  14. ],
                [ 15. ,  16. ,  17. ,  18. ,  19. ],
                [ 20. ,  21. ,  22. ,  23. ,  24. ]])
```

In preceding code, we changed the value of the first five elements to nan by putting a condition with index. x[x<5] refers to the elements which indexed for 0, 1, 2, 3, and 4. Then we overwrite these values with the mean of each column(excluding nan values). There are many other useful methods in array operations in order help your code be more concise:

Method	Description
np.concatenate	Join to the matrix in a sequence with a given matrix
np.repeat	Repeat the element of an array along a specific axis
np.delete	Return a new array with the deleted subarrays
np.insert	Insert values before the specified axis
np.unique	Find unique values in an array
np.tile	Create an array by repeating a given input for a given number of repetitions

Indexing, slicing, reshaping, resizing, and broadcasting

When you are working with huge arrays in machine learning projects, you often need to index, slice, reshape, and resize.

Indexing is a fundamental term used in mathematics and computer science. As a general term, indexing helps you to specify how to return desired elements of various data structures. The following example shows indexing for a list and a tuple:

```
In [74]: x = ["USA","France", "Germany","England"]
         x[2]
Out[74]: 'Germany'
In [75]: x = ('USA',3,"France",4)
         x[2]
Out[75]: 'France'
```

In NumPy, the main usage of indexing is controlling and manipulating the elements of arrays. It's a way of creating generic lookup values. Indexing contains three child operations, which are field access, basic slicing, and advanced indexing. In field access, you just specify the index of an element in an array to return the value for a given index.

NumPy is very powerful when it comes to indexing and slicing. In many cases, you need to refer your desired element in an array and do the operations on this sliced area. You can index your array similarly to what you do with tuples or lists with square bracket notations. Let's start with field access and simple slicing with one-dimensional arrays and move on to more advanced techniques:

```
In [76]: x = np.arange(10)
         x
Out[76]: array([0, 1, 2, 3, 4, 5, 6, 7, 8, 9])
In [77]: x[5]
Out[77]: 5
In [78]: x[-2]
Out[78]: 8
In [79]: x[2:8]
Out[79]: array([2, 3, 4, 5, 6, 7])
In [80]: x[:]
Out[80]: array([0, 1, 2, 3, 4, 5, 6, 7, 8, 9])
In [81]: x[2:8:2]
Out[81]: array([2, 4, 6])
```

Indexing starts from 0, so when you create an array with an element, your first element is indexed as x[0], the same way as your last element, x[n-1]. As you can see in the preceding example, x[5] refers to the sixth element. You can also use negative values in indexing. NumPy understands these values as the nth orders backwards. Like in the example, x[-2] refers to the second to last element. You can also select multiple elements in your array by stating the starting and ending indexes and also creating sequential indexing by stating the increment level as a third argument, as in the last line of the code.

So far, we have seen indexing and slicing in 1D arrays. The logic does not change, but for the sake of demonstration, let's do some practice for multidimensional arrays as well. The only thing that changes when you have multidimensional arrays is just having more axis. You can slice the n-dimensional array as [slicing in x-axis, slicing in y-axis] in the following code:

```
In [82]: x = np.reshape(np.arange(16),(4,4))
         x
Out[82]: array([[ 0,  1,  2,  3],
                [ 4,  5,  6,  7],
                [ 8,  9, 10, 11],
                [12, 13, 14, 15]])
In [83]: x[1:3]
Out[83]: array([[ 4,  5,  6,  7],
                [ 8,  9, 10, 11]])
In [84]: x[:,1:3]
Out[84]: array([[ 1,  2],
                [ 5,  6],
                [ 9, 10],
                [13, 14]])
In [85]: x[1:3,1:3]
Out[85]: array([[ 5,  6],
                [ 9, 10]])
```

You sliced the arrays row and column-wise, but you haven't sliced the elements in a more irregular or more dynamic fashion, which means you always slice them in a rectangular or square way. Imagine a 4*4 array that we want to slice as follows:

$$X = \begin{bmatrix} X_{11} & X_{12} & X_{13} & X_{14} \\ X_{21} & X_{22} & X_{23} & X_{24} \\ X_{31} & X_{32} & X_{33} & X_{34} \\ X_{41} & X_{42} & X_{43} & X_{41} \end{bmatrix}$$

To obtain the preceding slicing, we execute the following code:

```
In [86]: x = np.reshape(np.arange(16),(4,4))
         x
Out[86]: array([[ 0,  1,  2,  3],
                [ 4,  5,  6,  7],
                [ 8,  9, 10, 11],
                [12, 13, 14, 15]])
In [87]: x[[0,1,2],[0,1,3]]
Out[87]: array([ 0,  5, 11])
```

In advanced indexing, the first part indicates the index of rows to be sliced and the second part indicates the corresponding columns. In the preceding example, you first sliced the 1st, 2nd, and 3rd rows ([0,1,2]) and then sliced the 1st, 2nd and 4th columns ([0,1,3]) into sliced rows.

The reshape and resize methods may seem similar, but there are differences in the outputs of these operations. When you reshape the array, it's just the output that changes the shape of the array temporarily, but it does not change the array itself. When you resize the array, it changes the size of the array permanently, and if the new array's size is bigger than the old one, the new array elements will be filled with repeated copies of the old ones. On the contrary, if the new array is smaller, a new array will take the elements from the old array with the order of index which is required to fill the new one. Please note that same data can be shared by different ndarrays which means that an ndarray can be a view to another ndarray. In such cases changes made in one array will have consequences on other views.

The following code gives an example of how the new array elements are filled when the size is bigger or smaller than the original array:

```
In [88]: x = np.arange(16).reshape(4,4)
         x
Out[88]: array([[ 0,  1,  2,  3],
                [ 4,  5,  6,  7],
                [ 8,  9, 10, 11],
                [12, 13, 14, 15]])
In [89]: np.resize(x,(2,2))
Out[89]: array([[0, 1],
                [2, 3]])
In [90]: np.resize(x,(6,6))
Out[90]: array([[ 0,  1,  2,  3,  4,  5],
                [ 6,  7,  8,  9, 10, 11],
                [12, 13, 14, 15,  0,  1],
                [ 2,  3,  4,  5,  6,  7],
                [ 8,  9, 10, 11, 12, 13],
                [14, 15,  0,  1,  2,  3]])
```

The last important term of this subsection is broadcasting, which explains how NumPy behaves in arithmetic operations of the array when they have different shapes. NumPy has two rules for broadcasting: either the dimensions of the arrays are equal, or one of them is 1. If one of these conditions is not met, then you will get one of the two errors: `frames are not aligned` or `operands could not be broadcast together`:

```
In [91]: x = np.arange(16).reshape(4,4)
         y = np.arange(6).reshape(2,3)
         x+y
         ----------------------------------------------------------
         ------------
         ValueError Traceback (most recent call last)
         <ipython-input-102-083fc792f8d9> in <module>()
         1 x = np.arange(16).reshape(4,4)
         2 y = np.arange(6).reshape(2,3)
         ----> 3 x+y
         12
         ValueError: operands could not be broadcast together with
shapes (4,4) (2,3)
```

You might have seen that you can multiply two matrices with shapes *(4, 4)* and *(4,)* or with *(2, 2)* and *(2, 1)*. The first case meets the condition of having one dimension so that the multiplication becomes a vector * array, which does not cause any broadcasting problems:

```
In [92]: x = np.ones(16).reshape(4,4)
         y = np.arange(4)
         x*y
Out[92]: array([[ 0., 1., 2., 3.],
                [ 0., 1., 2., 3.],
                [ 0., 1., 2., 3.],
                [ 0., 1., 2., 3.]])
In [93]: x = np.arange(4).reshape(2,2)
         x
Out[93]: array([[0, 1],
                [2, 3]])
In [94]: y = np.arange(2).reshape(1,2)
         y
Out[94]: array([[0, 1]])
In [95]: x*y
Out[95]: array([[0, 1],
                [0, 3]])
```

The preceding code block gives an example for the second case, where during computation small arrays iterate through the large array and the output is stretched across the whole array. That's the reason why there are *(4, 4)* and *(2, 2)* outputs: during the multiplication, both arrays are broadcast to larger dimensions.

Summary

In this chapter, you got familiar with NumPy basics for array operations and refreshed your knowledge about basic matrix operations. NumPy is an extremely important library for Python scientific stacks, with its extensive methods for array operations. You have learned how to work with multidimensional arrays and covered important topics such as indexing, slicing, reshaping, resizing, and broadcasting. The main goal of this chapter was to give you a brief idea of how NumPy works when it comes to numerical datasets, which will be helpful in your daily data analysis work.

In the next chapter, you will learn the basics of linear algebra and complete practical examples with NumPy.

Linear Algebra with NumPy 2

One of the major divisions of mathematics is **algebra**, and linear algebra in particular, focuses on linear equations and mapping linear spaces, namely vector spaces. When we create a linear map between vector spaces, we are actually creating a data structure called a matrix. The main usage of linear algebra is to solve simultaneous linear equations, but it can also be used for approximations for non-linear systems. Imagine a complex model or system that you are trying to understand, think of it as a non-linear model. In such cases, you can reduce the complex, non-linear characteristics of the problem into simultaneous linear equations, and you can solve them with the help of linear algebra.

In computer science, linear algebra is heavily used in **machine learning** (**ML**) applications. In ML applications, you deal with high-dimensional arrays, which can easily be turned into linear equations where you can analyze the interaction of features in a given space. Imagine a case where you are working on an image recognition project and your task is to detect a tumor in the brain from MRI images. Technically, your algorithm should act like a doctor, where it scans the given input and detects the tumor in the brain. A doctor has the advantage of being able to spot anomalies; the human brain has been evolving through thousands of years to interpret visual input. Without much effort, a human can capture anomalies intuitively. However, for an algorithm to perform a similar task, you should think about this process in as much detail as possible to understand how you can formally express it so that the machines can understand.

First, you should think about how MRI data is stored in a computer, which processes only 0s and 1s. The computer actually stores pixel intensities in structures, called matrices. In other words, you will convert an MRI as a vector of dimensions, **N2**, where each element consists of pixel values. If this MRI has a 512 x 512 dimension, each pixel will be one point in 262,144 in pixels. Therefore, any computational manipulation that you will do in this Matrix would most likely use Linear Algebra principles. If this example is not enough to demonstrate the importance of Linear Algebra in ML, then let's look at a popular example in deep learning. In a nutshell, deep learning is an algorithm that uses neural network structure to learn the desired output (label) by continuously updating the weights of neuron connections between layers. A graphical representation of a simple deep learning algorithm is as follows:

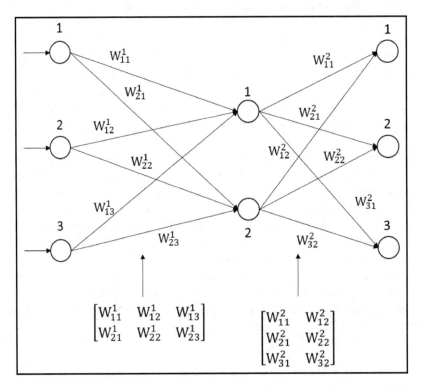

Neural networks store weights between layers and bias values in matrices. As these are the parameters that you try to tune in to your deep learning model in order to minimize your loss function, you continuously make computations and update them. In general, ML models require heavy calculations and need to be trained for big datasets to provide efficient results. This is why linear algebra is a fundamental part of ML.

In this chapter, we will use numpy library but note that most linear algebra functions are also imported by scipy and they are more properly belong to it. Ideally, in most cases, you import both of these libraries and perform computations. One important feature of scipy is to have fully-featured versions of the linear algebra modules. We highly encourage you to review scipy documentation and practice using same operations with scipy throughout this chapter. Here's the link for linear algebra module of scipy: `https://docs.scipy.org/doc/scipy/reference/linalg.html`.

In this chapter, we will cover the following topics:

- Vector and matrix mathematics
- What's an eigenvalue and how do we compute it?
- Computing the norm and determinant
- Solving linear equations
- Computing gradient

Vector and matrix mathematics

In the previous chapter, you practiced introductory operations with vectors and matrices. In this section, you will practice more advanced vector and matrix operations that are heavily used in linear algebra. Let's remember the dot product perspective on matrix manipulation and how it can be done with different methods when you have 2-D arrays. The following code block shows alternative ways of performing dot product calculation:

```
In [1]: import numpy as np
        a = np.arange(12).reshape(3,2)
        b = np.arange(15).reshape(2,5)
        print(a)
        print(b)
Out[1]:
[[ 0  1]
 [ 2  3]
 [ 4  5]]
[[ 0  1  2  3  4]
 [ 5  6  7  8  9]]
In [2]: np.dot(a,b)
Out[2]: array([[ 5,  6,  7,  8,  9],
               [15, 20, 25, 30, 35],
               [25, 34, 43, 52, 61]])
In [3]: np.matmul(a,b)
Out[3]: array([[ 5,  6,  7,  8,  9],
```

```
              [15,  20,  25,  30,  35],
              [25,  34,  43,  52,  61]]])
In [4]: a@b
Out[4]: array([[ 5,   6,   7,   8,   9],
              [15,  20,  25,  30,  35],
              [25,  34,  43,  52,  61]]])
```

Inner products and dot products are very important in ML algorithms such as supervised learning. Let's get back to our example about tumor detection. Imagine we have three images (MRIs): the first with a tumor (A), the second without a tumor (B) and the third one an unknown MRI that you want to label as *with tumor* or *without tumor*. The following graph shows a geometric representation of a dot product for vector a and b:

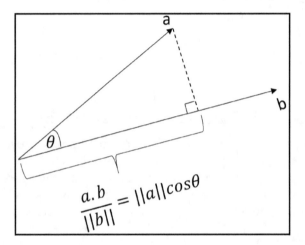

As a very simple example, the dot product will show you the similarity between these two vectors, so if your unknown MRI's direction is close to vector A, then your algorithm will classify as MRI with tumor. Otherwise, it will classify as without tumor. If you want to multiply two or more arrays in a single function, `linalg.multi_dot()` is very convenient. Another way to do this operation is by multiplying the arrays nested by `np.dot()` but you need to know which computation order will be fastest because in `linalg.multi_dot()` this optimization occurs automatically. The following code block does the same dot product operations, with different methods, as we mentioned previously:

```
In [5]: from numpy.linalg import multi_dot
        a = np.arange(12).reshape(4,3)
        b = np.arange(15).reshape(3,5)
        c = np.arange(25).reshape(5,5)
        multi_dot([a, b, c])
```

```
Out[5]: array([[ 1700,  1855,  2010,  2165,  2320],
               [ 5300,  5770,  6240,  6710,  7180],
               [ 8900,  9685, 10470, 11255, 12040],
               [12500, 13600, 14700, 15800, 16900]])
In [6]: a.dot(b).dot(c)
Out[6]: array([[ 1700,  1855,  2010,  2165,  2320],
               [ 5300,  5770,  6240,  6710,  7180],
               [ 8900,  9685, 10470, 11255, 12040],
               [12500, 13600, 14700, 15800, 16900]])
```

As you see in the following code, the multi_dot() method decreases running time by 40% even with three small matrices. This time gap can increase tremendously if the matrix amount and size increase. Consequently, you should use the multi_dot() method in order to be sure of the fastest evaluation order. The following code will compare the execution time of these two methods:

```
In [7]: import numpy as np
        from numpy.linalg import multi_dot
        import time
        a = np.arange(120000).reshape(400,300)
        b = np.arange(150000).reshape(300,500)
        c = np.arange(200000).reshape(500,400)
        start = time.time()
        multi_dot([a,b,c])
        ft = time.time()-start
        print ('Multi_dot tooks', time.time()-start,'seconds.')
        start_ft = time.time()
        a.dot(b).dot(c)
        print ('Chain dot tooks', time.time()-start_ft,'seconds.')
Out[7]:
Multi_dot tooks 0.14687418937683105 seconds.
Chain dot tooks 0.1572890281677246 seconds.
```

There are two more important methods in NumPy's linear algebra library, namely outer() and inner(). The outer method computes the outer product of two vectors. The inner method behaves differently depending on the arguments it takes. If you have two vectors as arguments, it produces an ordinary dot product, but when you have a higher dimensional array, it returns the sum product over the last axes as similarly in tensordot(). You will see tensordot() later in this chapter. Now, let's focus on the outer() and inner() methods first. The following example will help you to understand what these functions do:

```
In [8]: a = np.arange(9).reshape(3,3)
        b = np.arange(3)
        print(a)
        print(b)
```

```
Out[8]:
[[0 1 2]
 [3 4 5]
 [6 7 8]]
[0 1 2]
In [9]: np.inner(a,b)
Out[9]: array([ 5, 14, 23])
In [10]: np.outer(a,b)
Out[10]: array([[ 0,  0,  0],
                [ 0,  1,  2],
                [ 0,  2,  4],
                [ 0,  3,  6],
                [ 0,  4,  8],
                [ 0,  5, 10],
                [ 0,  6, 12],
                [ 0,  7, 14],
                [ 0,  8, 16]])
```

In the preceding example for the inner() method, the ith row of the array produces a scalar product with the vector and the sum becomes the ith element of the output array, so the output array is constructed as follows:

$$[0x0+1x1+2x2, 0x3+1x4 +2x5, 0x6 +1x7+2x8] = [5, 14, 23]$$

In the following code block, we perform the outer() method for the same array but with one-dimension. As you noticed the result is exactly the same as we do with 2-D array:

```
In [11]: a = np.arange(9)
         np.ndim(a)
Out[11]: 1
In [12]: np.outer(a,b)
Out[12]: array([[ 0,  0,  0],
                [ 0,  1,  2],
                [ 0,  2,  4],
                [ 0,  3,  6],
                [ 0,  4,  8],
                [ 0,  5, 10],
                [ 0,  6, 12],
                [ 0,  7, 14],
                [ 0,  8, 16]])
```

The outer() method computes the outer product of vectors, in our example, a 2-D array. That does not change the functionality of the method but flattens the 2-D array to a vector and does the computation.

Before moving to decomposition, the last thing this subsection covers is the tensordot() method. In a data science project, we mostly work with n-dimensional data which you need to do discovery and apply ML algorithms. Previously, you learned about vectors and matrices. A tensor is a generic mathematical object of vectors and matrices that can keep the relationships of vectors in high-dimensional spaces.

The tensordot() method is used for the contraction of two tensors; in other words, it reduces the dimensionality (tensor order) by summing the products of two tensors over the specified axes. The following code block shows an example of tensordot() operation for two arrays:

```
In [13]: a = np.arange(12).reshape(2,3,2)
         b = np.arange(48).reshape(3,2,8)
         c = np.tensordot(a,b, axes =([1,0],[0,1]))
         print(a)
         print(b)
Out[13]:
[[[ 0  1]
  [ 2  3]
  [ 4  5]]

 [[ 6  7]
  [ 8  9]
  [10 11]]]
[[[ 0  1  2  3  4  5  6  7]
  [ 8  9 10 11 12 13 14 15]]

 [[16 17 18 19 20 21 22 23]
  [24 25 26 27 28 29 30 31]]

 [[32 33 34 35 36 37 38 39]
  [40 41 42 43 44 45 46 47]]]
In [14]: c
Out[14]: array([[ 800,  830,  860,  890,  920,  950,  980, 1010],
                [ 920,  956,  992, 1028, 1064, 1100, 1136, 1172]])
```

What's an eigenvalue and how do we compute it?

An eigenvalue is a coefficient of an eigenvector. By definition, an **eigenvector** is a non zero vector that only changes by a scalar factor when linear transformation is applied. In general, when linear transformation is applied to a vector, its span (the line passing through its origin) is shifted, but some special vectors are not affected by these linear transformations and remain on their own span. These are what we call eigenvectors. The linear transformation affects them only by stretching or squishing them as you are multiplying this vector with a scalar. The value of this scalar is called the eigenvalue. Let's say we have a matrix A, which will be used in linear transformation. We can represent the eigenvalue and eigenvector in a mathematical statements as follows:

$$Av = \lambda v$$

Here, v is the eigenvector and λ denotes the eigenvalue. In the left part of the equation, the vector is transformed by a matrix and the result is just a scalar version of the same vector. In addition, notice that the left-hand side is actually a matrix-vector multiplication but the right-hand side is a scalar vector multiplication. In order to make the multiplication type for sides the same, we can multiply the λ with the identity matrix, which will not change the right-hand side. But change λ from scalar to matrix, and both sides will be a matrix- vector multiplication:

$$Av = (\lambda I)v$$

If we subtract the right-hand side and factor out v you will have the following equation:

$$(A - \lambda I)v = 0$$

Therefore, our new matrix will be as follows:

$$\begin{bmatrix} x_{11} - \lambda & x_{12} & x_{13} \\ x_{21} & x_{22} - \lambda & x_{23} \\ x_{31} & x_{32} & x_{33} - \lambda \end{bmatrix}$$

As you will already know, you are trying to calculate the eigenvalue for the eigenvector other than 0; (if v=0, that doesn't make sense) so the eigenvector (v) cannot be 0. Therefore, you are trying to solve the equation for the following:

$$det(A - \lambda I) = 0$$

The preceding formula is calculating the determinant of a matrix, as there is only one condition where a non-zero matrix is equal to 0, which is when its determinant equals zero. Having a zero determinant for a given matrix means that the transformation with that matrix squishes everything into a smaller dimension. You will see determinants in more detail in the next section. The real aim here is to find the λ that makes the determinant zero and squishes the space into a lower dimension. After finding λ, we can calculate the eigenvector (v) from the following equation:

$$Av = \lambda v$$

In ML algorithms, you work with large numbers of dimensions. The main problem is not having a huge dimension, but your algorithm's compatibility and performance with them. For example, in **PCA (principal component analysis)**, you try to discover the most meaningful linear combination of your dimension. The main idea in PCA is reducing the dimension of your dataset while minimizing the loss of information. The loss of information here is actually variance of your features. Let's say you have five features and a class label for five samples as follows:

Feature 1	Feature 2	Feature 3	Feature 4	Feature 5	Class
1.45	42	54	1.001	1.05	Dog
2	12	34	1.004	24.1	Cat
4	54	10	1.004	13.4	Dog
1.2	31	1	1.003	42.1	Cat
5	4	41	1.003	41.4	Dog

In the preceding table, in **Feature 4,** it actually makes no significant difference whether the class label is **Dog** or **Cat**. This feature will be redundant in my analysis. The main goal here is to keep the features that strongly differ across the classes, so the feature value plays an important role in my decision.

In the following example, we will use an additional library which is called `scikit-learn` in order to load the dataset and perform PCA by using this library as well. Scikit-learn is a free machine learning library for python which is built on top of SciPy. It has many built-in functions and modules for many machine learning algorithms such as classification, regression, clustering. SVM, DBSCAN, k-means and many others.

Scikit-learn has a great compatibility with numpy, scipy and pandas. You can use numpy array in `sklearn` as a data structure. Moreover, you can also use `sklearn_pandas` to transform scikit-learn pipeline outputs to pandas dataframe. Scikit-learn also support automating machine learning algorithms with the help of `auto_ml` and `auto-sklearn` libraries. `sklearn` is a way of typing scikit-learn name in python. That's why you use `sklearn` when you import skicit-learn and using functions.

Now it's time to do a practical exercise in order to see the usage and importance of eigenvalues and eigenvectors in PCA. In the following code, you import and apply PCA with NumPy, then you will compare your results with `sklearn` is built-in methods for validation. You will use a breast-cancer dataset from the sklearn datasets library. Let's import the required libraries and the dataset first, then you need to standardize your data. Standardization is very important, and sometimes it's even a requirement for estimators in some ML libraries such as scikit-learn. In our example, the `StandardScaler()` method is used in order to standardize the features; the main purpose here is to have features as in standard normally distributed data (mean = 0 and unit variance). The `fit_transform()` method is used to transform original data in a form that the distribution will have a mean value 0 and standard deviation 1. It will calculate the required parameters and apply the transformation. As we use `StandardScaler()` this parameters will be μ and σ. Please keep in mind that standardization does not produce a normally distributed data from our original dataset. It's just rescaled the data where we have a mean of zero and standard deviation one:

```
In [15]: import numpy as np
         from sklearn import decomposition, datasets
         from sklearn.preprocessing import StandardScaler
         data = datasets.load_breast_cancer()
         cancer = data.data
         cancer = StandardScaler().fit_transform(cancer)
         cancer.shape
Out[15]: (569, 30)
```

As you checked the shape of your data in the preceding code, it shows the data consists of 569 rows and 30 columns. You can also compare the data in the beginning and after standardization just to understand more clearly how the original data was transformed. Following code block shows you an example of a transformation of a column in cancer data:

```
In [16]: before_transformation = data.data
         before_transformation[:10,:1]
Out[16]: array([[ 17.99],
                [ 20.57],
                [ 19.69],
```

```
                      [ 11.42],
                      [ 20.29],
                      [ 12.45],
                      [ 18.25],
                      [ 13.71],
                      [ 13. ],
                      [ 12.46]]])
In [17]: cancer[:10,:1]
Out[17]: array([[ 1.09706398],
                      [ 1.82982061],
                      [ 1.57988811],
                      [-0.76890929],
                      [ 1.75029663],
                      [-0.47637467],
                      [ 1.17090767],
                      [-0.11851678],
                      [-0.32016686],
                      [-0.47353452]])
```

As you see from the results, the original data has been transformed into a form of standardized version. This is calculated by the following formula:

$$Z = \frac{x - \mu}{\sigma}$$

- x= The value that is being standardized (values in original data)
- μ= The mean of the distribution
- σ= Standard deviation of the distribution
- Z= Standardized values

After transforming the data, you will calculate the covariance matrix as follows in order to calculate eigenvalue and eigenvector with the `np.linalg.eig()` method, then use them in decomposition:

```
In [18]: covariance_matrix = np.cov(cancer,rowvar=False)
         covariance_matrix.shape
Out[18]: (30, 30)
In [19]: eig_val_cov, eig_vec_cov = np.linalg.eig(covariance_matrix)
         eig_pairs = [(np.abs(eig_val_cov[i]), eig_vec_cov[:,i]) for i in
         range(len(eig_val_cov))]
```

As you will have noticed in the preceding code, we calculate the covariance matrix constructed for all features. As we have 30 features in the dataset, the covariance matrix has a shape of *(30,30)* 2-D array. The following code block sorts the eigenvalues in descending order:

```
In [20]: sorted_pairs = eig_pairs.sort(key=lambda x: x[0], reverse=True)
         for i in eig_pairs:
             print(i[0])
Out[20]:
13.3049907944
5.70137460373
2.82291015501
1.98412751773
1.65163324233
1.2094822398
0.676408881701
0.47745625469
0.417628782108
0.351310874882
0.294433153491
0.261621161366
0.241782421328
0.157286149218
0.0943006956011
0.0800034044774
0.0595036135304
0.0527114222101
0.049564700213
0.0312142605531
0.0300256630904
0.0274877113389
0.0243836913546
0.0180867939843
0.0155085271344
0.00819203711761
0.00691261257918
0.0015921360012
0.000750121412719
0.000133279056663
```

You need to sort the eigenvalues in decreasing order to decide which eigenvector(s) you want to drop in order to lower the dimensional workspace. As you've seen in the preceding sorted list, the first two eigenvectors with high eigenvalues bear the most information about the distribution of the data; therefore the rest will be dropped for lower-dimensional subspace:

```
In [21]: matrix_w = np.hstack((eig_pairs[0][1].reshape(30,1), eig_pairs[1]
[1].reshape(30,1)))
         matrix_w.shape
         transformed = matrix_w.T.dot(cancer.T)
         transformed = transformed.T
         transformed[0]
Out[21]: array([ 9.19283683, 1.94858307])
In [22]: transformed.shape
Out[22]: (569, 2)
```

In the preceding code block, the first two eigenvectors stacked horizontally as they will be used in matrix multiplication to transform our data for the new dimensions onto the new subspace. The final data transformed from (569, 30) to (569, 2) matrix, which means that 28 features dropped off during the PCA process:

```
In [23]: import numpy as np
         from sklearn import decomposition
         from sklearn import datasets
         from sklearn.preprocessing import StandardScaler
         pca = decomposition.PCA(n_components=2)
         x_std = StandardScaler().fit_transform(cancer)
         pca.fit_transform(x_std)[0]
Out[23]: array([ 9.19283683, 1.94858307])
```

On the other hand, there are built-in functions in other libraries that do the same operations that you do. In scikit-learn, there are many built-in methods that you can use for ML algorithms. As you see in the preceding code block, the same PCA was performed by three lines of code with two methods. Nevertheless, the intention of this example is to show you the importance of eigenvalues and eigenvectors therefore the book shows the plain vanilla way of PCA with NumPy.

Computing the norm and determinant

This subsection will introduce two important values in linear algebra, namely the norm and determinant. Briefly, the norm gives length of a vector. The most commonly used norm is the L^2-norm, which is also known as the Euclidean norm. Formally, the L^p-norm of x is calculated as follows:

$$||x||_p = \sqrt[p]{\sum_i |x_i|^p}$$

The L^0-norm is actually the cardinality of a vector. You can calculate it by just counting the total number of non-zero elements. For example, the vector $A =[2,5,9,0]$ contains three non-zero elements, therefore $||A||_0 = 3$. The following code block shows the same norm calculation with numpy:

```
In [24]: import numpy as np
         x = np.array([2,5,9,0])
         np.linalg.norm(x,ord=0)
Out[24]: 3.0
```

In NumPy, you can calculate the norm of the vector with the use of the `linalg.norm()` method. The first parameter is the input array and the `ord` parameter is for order of the norm. The L^1-norm is also known as *Taxicab norm* or *Manhattan norm*. It calculates the length of the vectors by calculating the rectilinear distances, therefore $||A||_1=(2+5+9)$ so $||A||_1 = 16$. The following code block shows the same norm calculation with numpy:

```
In [25]: np.linalg.norm(x,ord=1)
Out[25]: 16.0
```

One of the uses of the L^1- norm is for the calculation of **mean-absolute error** (**MAE**) as in the following formula:

$$MAE(x_1, x_2) = \frac{1}{n}||x_1 - x_2||_1$$

The L^2-norm is the most popular norm. It calculates the length of the vector by applying the Pythagorean theorem, therefore $||A||_2 = \sqrt{2^2 + 5^2 + 9^2}$, so $||A||_2 = 10.48$:

```
In [26]: np.linalg.norm(x,ord=2)
Out[26]: 10.488088481701515
```

One of the well-known applications of the L^2-norm is the **mean-squared error (MSE)** calculations. As a matrix consists of vectors, similarly, the norm gives the length or size, but the interpretation and computation is slightly different. In previous chapters, you learned about matrix multiplication with vectors. When you multiply the matrix with a vector in the result, you stretch the vector. The norm of the matrix reveals how much that matrix could possibly stretch a vector. Let's see how the $L^1$and $L\infty$-norm is calculated for matrix A as follows:

$$A = \begin{bmatrix} 3 & 7 & 6 \\ -2 & -5 & 4 \\ 1 & 3 & -14 \end{bmatrix}$$

Let's assume that you have an $m \times n$ matrix. The calculation for $||A||_1$ is as follows:

$$||A||_1 = Max_{1 \leq j \leq n} \sum_{i=0}^{m} |a_{ij}|,$$

So the result will be as follows:

$||A||_1 = max(3+|-2|+1; 7+|-5|+3; 6+4+|-14|) = max(6,15,24) = 24$

Let's calculate the norm of the same array with NumPy using the `linalg.norm()` method as follows:

```
In [27]: a = np.array([3,7,6,-2,-5,4,1,3,-14]).reshape(3,3)
In [28]: a
Out[28]: array([[ 3,  7,  6],
                [ -2,  -5,  4],
                [ 1,  3,  -14]])
In [29]: np.linalg.norm(a, ord=1)
Out[29]: 24.0
In [30]: np.linalg.norm(a, np.inf)
Out[30]: 18.0
```

In the preceding calculation $||A||_1$ has calculated the max element as column-wise first and gives the maximum result in all columns, which will be the norm of the matrix for the first order. For $L\infty$, the max element calculations perform row-wise and the give the maximum result in all rows that get the norm matrix for infinity order. The calculation looks as follows:

$||A||\infty = max(3+7+6; |-2|+|-5|+4; 1+3+|-14|) = max(16,11,18) = 18$

In order to validate and use NumPy functionality, the same calculations are done with the `linalg.norm()` method as we do in vector norm calculations. Calculations for the first and infinite order are relatively more straightforward than the general calculation for the *p*-norm (where *p*>2), such as the Euclidean/Frobenius norm (where *p*=2). The following formula shows the formal formula for the p-norm, and just by replacing the *p* with 2 you can formulate the Euclidean/Frobenius norm for a given array:

$$||A||_p = (\sum_{i=1}^{m} \sum_{j=0}^{n} a_{ij}^p)^{1/p}$$

For the special case of *p*=2, this becomes:

$$||A||_2 = (\sum_{i=1}^{m} \sum_{j=0}^{n} a_{ij}^2)^{1/2}$$

The following code block shows L-2 norm calculation for array a:

```
In [31]: np.linalg.norm(a, ord=2)
Out[31]: 15.832501006406099
```

In NumPy, the Euclidean/ Frobenius norm can be calculated by setting the order parameter to 2 in `linalg.norm()` method as you see in the preceding code. In ML algorithms, the norm is heavily used in distance calculation for the feature space. For example, in pattern recognition, norm calculations used in k-nearest neighbors algorithm (k-NN) in order to create distance metric for continuous variables or discrete variables. Similarly, norms are very important for distance metrics in k-means clustering as well. The most popular orders are the Manhattan norm and the Euclidean norm.

Another important concept in linear algebra is calculating the determinant of the matrices. By definition, the determinant is the scaling factor of a given matrix in linear transformation. In the previous section, while you are calculating eigenvalues and eigenvectors, you multiply a given matrix by an eigenvector and assume that their determinant will be zero. In another words, you assume that after you multiply the eigenvector by your matrix, you will flatten your matrix into a lower dimension. The determinant formula for a 2 x 2 and 3 x 3 matrix is as follows:

$$|A| = \begin{vmatrix} a & b \\ c & d \end{vmatrix} = ad - bc$$

$$|A| = \begin{vmatrix} a & b & c \\ d & e & f \\ g & h & i \end{vmatrix} = a\begin{vmatrix} e & f \\ h & i \end{vmatrix} + b\begin{vmatrix} d & f \\ g & i \end{vmatrix} + c\begin{vmatrix} d & e \\ g & h \end{vmatrix}$$

$$= a(ei - fh) + b(fg - di) + c(dh - eg)$$

In NumPy, you can use the `linalg.det()` method to calculate the determinant of your matrix. Let's calculate an example of the *2 x 2* and *3 x 3* matrix determinant with the preceding formula and cross-validate our result with NumPy:

$$|A| = \begin{vmatrix} 2 & 3 \\ 1 & 4 \end{vmatrix}$$

So the calculation will be as follows:

$$det(A) = (2 \times 4) - (3 \times 1) = 5$$

For the case where we have *3 x 3* matrix, this becomes:

$$|B| = \begin{vmatrix} 2 & 3 & 5 \\ 1 & 4 & 8 \\ 5 & 6 & 2 \end{vmatrix}$$

$$det(B) = (2 \times (8 - 48)) + (3 \times (40 - 2)) + (5 \times (6 - 20))$$

$$det(B) = -80 + 114 - 70 = -36$$

Let's calculate the determinant of the same arrays with NumPy using the `linalg.det()` method as follows:

```
In [32]: A= np.array([2,3,1,4]).reshape(2,2)
In [33]: A
Out[33]: array([[2, 3],
                [1, 4]])
In [34]: np.linalg.det(A)
Out[34]: 5.0
In [35]: B= np.array([2,3,5,1,4,8,5,6,2]).reshape(3,3)
In [36]: B
Out[36]: array([[2, 3, 5],
                [1, 4, 8],
                [5, 6, 2]])
In [37]: np.linalg.det(B)
Out[37]: -36.0
```

The determinant of a transformation actually shows the factor of how much the volume will extend or compress. If the determinant of a matrix *A* equals 2, that means the transformation of this matrix will extend the volume by 2. If you are doing a chain multiplication, you can also calculate how much the volume changes after the transformation. Let's say you multiply two matrices, that means that there will be two transformations. If *det(A)* = 2 and *det(B)*=3, the total transformation factor will be multiplied by 6 as *det(AB)* = *det(A)det(B)*.

Lastly, this chapter will cover a very useful value for your ML models, the value called **trace**. By definition, the trace is the sum of diagonal elements of a matrix. In ML models, in most cases, you work with several regression models to explain your data. It's very likely that some of these models explain your data quality more or less the same, therefore in such cases, you always tend to move forward with the simpler model. Whenever you have to do this trade-off, the trace value becomes very valuable for quantifying the complexity. The following code block shows trace calculation with numpy for 2-D and 3-D matrices:

```
In [38]: a = np.arange(9).reshape(3,3)
In [39]: a
Out[39]: array([[0, 1, 2],
                [3, 4, 5],
                [6, 7, 8]])
In [40]: np.trace(a)
Out[40]: 12
In [41]: b = np.arange(27).reshape(3,3,3)
In [42]: b
Out[42]: array([[[ 0,  1,  2],
                 [ 3,  4,  5],
                 [ 6,  7,  8]],
                [[ 9, 10, 11],
                 [12, 13, 14],
                 [15, 16, 17]],

                [[18, 19, 20],
                 [21, 22, 23],
                 [24, 25, 26]]])

In [43]: np.trace(b)
Out[43]: array([36, 39, 42])
```

In NumPy, you can use the `trace()` method to calculate the trace value of your matrix. If your matrix is 2-D then the trace is the sum along the diagonal. If your matrix has a dimension more than 2, then the trace is an array of sums along the diagonals. In the preceding example, matrix *B* has three dimensions so the trace array is constructed as follows:

$$TraceArray = [0 + 12 + 24, 1 + 13 + 25, 2 + 14 + 26] = [36, 39, 42]$$

In this subsection, you learned how to calculate the norm, determinant, trace, and usage in ML algorithms. The main aim was to learn these concepts and become familiar with the NumPy linear algebra library.

Solving linear equations

In this section, you will learn how to solve linear equations by using the `linalg.solve()` method. When you have a linear equation to solve, as in the form $Ax = B$, in simple cases you can just calculate the inverse of A and then multiply it by B to get the solution, but when A has a high dimensionality, that makes it very hard computationally to calculate the inverse of A. Let's start with an example of three linear equations with three unknowns, as follows:

$$2a + b + 2c = 8$$

$$3a + 2b + c = 3$$

$$b + c = 4$$

So, these equations can be formalized as follows with matrices:

$$A = \begin{bmatrix} 2 & 1 & 2 \\ 3 & 2 & 1 \\ 0 & 1 & 1 \end{bmatrix}, x = \begin{bmatrix} a \\ b \\ c \end{bmatrix}, B = \begin{bmatrix} 8 \\ 3 \\ 4 \end{bmatrix}$$

Then, our problem is to solve $Ax = B$. We can calculate the solution with a plain vanilla NumPy without using `linalg.solve()`. After inverting the A matrix, you will multiply with B in order to get results for x. In the following code block, we calculate the dot product for the inverse matrix of A and B in order to calculate x:

$$x = A^{-1}B$$

```
In [44]: A = np.array([[2, 1, 2], [3, 2, 1], [0, 1, 1]])
         A
Out[44]: array([[2, 1, 2],
                [3, 2, 1],
                [0, 1, 1]])
In [45]: B = np.array([8,3,4])
```

```
          B
Out[45]: array([8,  3,  4])
In [46]: A_inv = np.linalg.inv(A)
         A_inv
Out[46]: array([[ 0.2,  0.2,  -0.6],
                [-0.6,  0.4,  0.8],
                [ 0.6,  -0.4,  0.2]])
In [47]: np.dot(A_inv,B)
Out[47]: array([-0.2,  -0.4,  4.4])
```

Finally, you get the result for *a =-0.2*, *b= -0.4* and *c=4.4*. Now, let's perform the same calculation with `linalg.solve()` as follows:

```
In [48]: A = np.array([[2, 1, 2], [3, 2, 1], [0, 1, 1]])
         B = np.array([8,3,4])
         x = np.linalg.solve(A, B)
         x
Out[48]: array([-0.2,  -0.4,  4.4])
```

In order to check our results, we can use the `allclose()` function, which is used to compare two arrays element-wise:

```
In [49]: np.allclose(np.dot(A, x), B)
Out[49]: True
```

Another important function for solving linear equations, which returns the least-square solution, is the `linalg.lstsq()` method. This function will return the parameters for the regression line. The main idea of the regression line is to minimize the sum of the squares of distance from each data point to the regression line. The sum of squares of distances actually quantify the total error of the regression line. Higher distance means higher error. As a result, we are looking for parameters that will minimize this error. In order to visualize our linear regression model, we will use a very popular 2-D plotting library for python which called `matplotlib`. The following code block runs the least-squares solution in our matrix and returns the weight and bias:

```
In [50]: from numpy import arange,array,ones,linalg
         from pylab import plot,show
         x = np.arange(1,11)
         A = np.vstack([x, np.ones(len(x))]).T
         A
Out[50]: array([[ 1.,  1.],
               [ 2.,  1.],
               [ 3.,  1.],
               [ 4.,  1.],
               [ 5.,  1.],
               [ 6.,  1.],
```

```
                    [ 7.,  1.],
                    [ 8.,  1.],
                    [ 9.,  1.],
                    [ 10.,  1.]])
In [51]: y = [5,  6,  6.5,  7,  8,9.5,  10,  10.4,13.1,15.5]
         w = linalg.lstsq(A,y)[0]
         w
Out[51]: array([ 1.05575758,  3.29333333])
In [52]: line = w[0]*x+w[1]
         plot(x,line,'r-',x,y,'o')
         show()
```

The output of the preceding code is as follows:

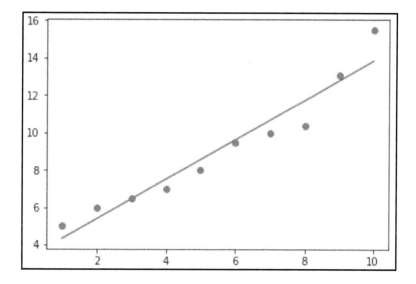

The preceding plot gives a result of regression line fitting which fits on our data. The model generates a linear line that can be used for prediction or forecasting. Although linear regressions have many assumptions (constant variance, independence of errors, linearity, and so on), it is still the most commonly used approach for modeling linear relationships between data points.

Computing gradient

When you have a linear line, you take the derivative so the derivative shows the slope of this line. Gradient is a generalization of the derivative when you have a multiple variable in your function, therefore the result of gradient is actually a vector function rather than a scalar value in derivative. The main goal of ML is actually finding the best model that fits your data. You can evaluate the meaning of the best as minimizing your loss function or objective function. Gradient is used for finding the value of the coefficients or a function that will minimize your loss or cost function. A well-known way of finding optimum points is taking the derivative of the objective function then setting it to zero to find your model coefficients. If you have more than one coefficient then it becomes a gradient rather than a derivative, and it becomes a vector equation rather than a scalar value. You can interpret a gradient as a vector at every point, which is directed to a next local minimum for your function. There is a very popular optimization technique for finding the minimum of a function by computing the gradient for each point and moving the coefficients along this direction, aiming to find the minimum. This method called **gradient descent**. In NumPy, the gradient() function is used to calculate the gradient of an array:

```
In [53]: a = np.array([1,3, 6, 7, 11, 14])
         gr = np.gradient(a)
         gr
Out[53]: array([ 2. , 2.5, 2. , 2.5, 3.5, 3. ])
```

Let's see how this gradient is calculated:

```
gr[0] = (a[1]-a[0])/1 = (3-1)/1 = 2
gr[1] = (a[2]-a[0])/2 = (6-1)/2 = 2.5
gr[2] = (a[3]-a[1])/2 = (7-3)/2 = 2
gr[3] = (a[4]-a[2])/2 = (11-6)/2 = 2.5
gr[4] = (a[5]-a[3])/2 = (14-7)/2 = 3.5
gr[5] = (a[5]-a[4])/1 = (14-11)/1 = 3
```

In the preceding code blocks, we calculated the gradient for a one-dimensional array. Let's add another dimension and see how the calculation changes as follows:

```
[54]: a = np.array([1,3, 6, 7, 11, 14]). reshape(2,3)
      gr = np.gradient(a)
      gr
Out[54]: [array([[ 6., 8., 8.],
                 [ 6., 8., 8.]]), array([[ 2. , 2.5, 3. ],
                 [ 4. , 3.5, 3. ]])]
```

In the case of a 2-D array, the gradient is calculated column-wise and row-wise as in the preceding code. Therefore, there will be a two-array return as a result; the first array stands for row direction and the second array is for column direction.

Summary

In this chapter, we covered vector and matrix operations for linear algebra. We looked at advanced matrix operations, especially featuring dot operations. You also learned about eigenvalues and eigenvectors and then practiced their use in **principal component analysis (PCA)**. Moreover, we covered the norm and determinant calculation and mentioned their importance and usage in ML. In the last two subsections, you learned how to convert linear equations into matrices and solve them, and looked at the computation and importance of gradients.

In the next chapter, we will use NumPy statistics to do explanatory data analysis to explore the 2015 United States Housing data.

3
Exploratory Data Analysis of Boston Housing Data with NumPy Statistics

Exploratory data analysis (**EDA**) is a crucial component of a data science project (as shown in Figure *Data Science Process*). Even though it is a very important step before applying any statistical model or machine learning algorithm to your data, it is often skipped or underestimated by many practitioners:

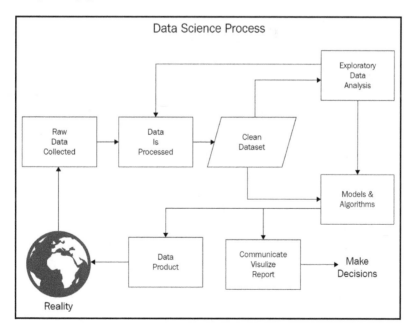

Data Science Process (https://en.wikipedia.org/wiki/Data_analysis)

John Wilder Tukey promoted exploratory data analysis in 1977 with his book *Exploratory Data Analysis*. In his book, he guides statisticians to analyze their datasets statistically by using several different visuals, which will help them to formulate their hypotheses. In addition, EDA is also used to prepare your analysis for advance modeling after you identify the key data characteristics and learn which questions you should ask about your data. At a high level, EDA is an assumption-free exploration of your data that takes advantage of quantitative methods, which allows you to visualize your results so that you can identify patterns, anomalies, and data characteristics. In this chapter, you will be using NumPy's built-in statistics methods to perform exploratory data analysis.

The following topics will be covered in this chapter:

- Loading and saving files
- Exploring the dataset
- Looking at the basic statistics
- Computing averages and variances
- Computing correlations
- Computing histograms

Loading and saving files

In this section, you will learn how to load/import your data and save it. There are many different ways of loading data, and the right way depends on your file type. You can load/import text files, SAS/Stata files, HDF5 files, and many others. **HDF** (**Hierarchical Data Format**) is one of the popular data formats which is used to store and organize large amounts of data and it is very useful while working with a multidimensional homogeneous arrays. For example, Pandas library has a very handy class named as `HDFStore` where you can easily work with HDF5 files. While working on data science projects, you will most likely see many of these types of files, but in this book, we will cover the most popular ones, such as **NumPy binary files, text files** (`.txt`), and **comma-separated values** (`.csv`) files.

If you have a large dataset in memory and on disk to manage, you can use a library called `bcolz`. It provides columnar and compressed data containers. bcolz objects are called `chunks` where you compressed whole data as bits then partially decompressed when you query. As data is compressed, it uses the storage very efficiently. `bcolz` objects improve data fetching performance as well. If you are interested more about the performance of this library. You can check query and speed comparison on their official GitHub repository `https://github.com/Blosc/bcolz/wiki/Query-Speed-and-Compression`.

When you are working with arrays, you will usually save them as NumPy binary files after you have finished working with them. The reason for this is that you need to store your array shape and data type as well. When you reload your array, you expect NumPy to remember it, and you can continue working from where you left off.
Moreover, NumPy binary files can store information about an array, even when you open the file on another machine with a different architecture. In NumPy, the `load()`, `save()`, `savez()`, and `savez_compressed()` methods help you to load and save NumPy binary files as follows:

```
In [1]: import numpy as np
In [2]: example_array = np.arange(12).reshape(3,4)
In [3]: example_array
Out[3]: array([[ 0,  1,  2,  3],
               [ 4,  5,  6,  7],
               [ 8,  9, 10, 11]])
In [4]: np.save('example.npy',example_array)
In [5]: d = np.load('example.npy')
In [6]: np.shape(d)
Out[6]: (3, 4)
In [7]: d
Out[7]: array([[ 0,  1,  2,  3],
               [ 4,  5,  6,  7],
               [ 8,  9, 10, 11]])
```

In the preceding code, we execute the following steps to a practice saving array as a binary file and how to load it back without affecting its shape:

1. Create an array with a shape *(3,4)*
2. Save the array as a binary file
3. Load back the array
4. Check whether the shape is still the same

Similarly, you can use the `savez()` function to save several arrays into a single file. If you want to save your files as compressed NumPy binary files, you can use `savez_compressed()` as follows:

```
In [8]: x = np.arange(10)
        y = np.arange(12)
        np.savez('second_example.npz',x, y)
        npzfile = np.load('second_example.npz')
        npzfile.files
Out[8]: ['arr_0', 'arr_1']
In [9]: npzfile['arr_0']
Out[9]: array([0, 1, 2, 3, 4, 5, 6, 7, 8, 9])
In [10]: npzfile['arr_1']
Out[10]: array([ 0, 1, 2, 3, 4, 5, 6, 7, 8, 9, 10, 11])
In [11]: np.savez_compressed('compressed_example.npz', first_array = x ,
second_array = y)
         npzfile = np.load('compressed_example.npz')
         npzfile.files
Out[11]: ['first_array', 'second_array']
In [12]: npzfile['first_array']
Out[12]: array([0, 1, 2, 3, 4, 5, 6, 7, 8, 9])
In [13]: npzfile['second_array']
Out[13]: array([ 0, 1, 2, 3, 4, 5, 6, 7, 8, 9, 10, 11])
```

When you save several arrays in a single file, if you give a keyword argument such as `first_array=x`, your array will be saved with this name. Otherwise, by default, your first array will be given a variable name, such as `arr_0`. NumPy has a built-in function called `loadtxt()` for loading data from a text file. Let's load a `.txt` file that consists of some integers and a header, which is a string at the top of the file:

```
In [14]: a = np.loadtxt "My_file.txt", delimiter='\t'

         ---------------------------------------------------------------------
         ValueError                        Traceback (most recent call last)
         <ipython-input-14-bde8ee3f2c6d> in <module>()
         ----> 1 a = np.loadtxt("My_file.txt", delimiter='\t')

         c:\users\mert_cuhadaroglu\appdata\local\programs\python\python36\lib\site-packages\numpy\lib\npyio.py in loadtxt(fname, dtype,
         comments, delimiter, converters, skiprows, usecols, unpack, ndmin, encoding)
            1090          # converting the data
            1091          X = None
         -> 1092          for x in read_data(_loadtxt_chunksize):
            1093              if X is None:
            1094                  X = np.array(x, dtype)

         c:\users\mert_cuhadaroglu\appdata\local\programs\python\python36\lib\site-packages\numpy\lib\npyio.py in read_data(chunk_size)
            1017
            1018          # Convert each value according to its column and store
         -> 1019          items = [conv(val) for (conv, val) in zip(converters, vals)]
            1020
            1021          # Then pack it according to the dtype's nesting

         c:\users\mert_cuhadaroglu\appdata\local\programs\python\python36\lib\site-packages\numpy\lib\npyio.py in <listcomp>(.0)
            1017
            1018          # Convert each value according to its column and store
         -> 1019          items = [conv(val) for (conv, val) in zip(converters, vals)]
            1020
            1021          # Then pack it according to the dtype's nesting

         c:\users\mert_cuhadaroglu\appdata\local\programs\python\python36\lib\site-packages\numpy\lib\npyio.py in floatconv(x)
             736          if '0x' in x:
             737              return float.fromhex(x)
         --> 738          return float(x)
             739
             740      typ = dtype.type

         ValueError: could not convert string to float: 'The following numbers are generated for the purporse of this chapter'
```

In the preceding code, you will get an error as you cannot convert a string to a float, which actually means that you are reading non-numeric values. The reason for this is that the file contains the string on top as a header as well as the numerical values. If you know how many lines your header has, you can skip these rows by using the skiprows parameter, as follows:

```
In [15]: a = np.loadtxt("My_file.txt", delimiter='\t', skiprows=4)
         a
Out[15]: array([  0.,   1.,   2.,   3.,   4.,   5.,   6.,   7.,   8.,
                  9.,  10.,  11.,  12.,  13.,  14.,  15.,  16.,  17.,
                 18.,  19.,  20.,  21.,  22.,  23.,  24.,  25.,  26.,
                 27.,  28.,  29.,  30.,  31.,  32.,  33.,  34.,  35.,
                 36.,  37.,  38.,  39.,  40.,  41.,  42.,  43.,  44.,
                 45.,  46.,  47.,  48.,  49.,  50.,  51.,  52.,  53.,
                 54.,  55.,  56.,  57.,  58.,  59.,  60.,  61.,  62.,
                 63.,  64.,  65.,  66.,  67.,  68.,  69.,  70.,  71.,
                 72.,  73.,  74.,  75.,  76.,  77.,  78.,  79.,  80.,
                 81.,  82.,  83.,  84.,  85.,  86.,  87.,  88.,  89.,
                 90.,  91.,  92.,  93.,  94.,  95.,  96.,  97.,  98.,
                 99., 100., 101., 102., 103., 104., 105., 106., 107.,
                108., 109., 110., 111., 112., 113., 114., 115., 116.,
                117., 118., 119., 120., 121., 122., 123., 124., 125.,
```

```
126., 127., 128., 129., 130., 131., 132., 133., 134.,
135., 136., 137., 138., 139., 140., 141., 142., 143.,
144., 145., 146., 147., 148., 149., 150., 151., 152.,
153., 154., 155., 156., 157., 158., 159., 160., 161.,
162., 163., 164., 165., 166., 167., 168., 169., 170.,
171., 172., 173., 174., 175., 176., 177., 178., 179.,
180., 181., 182., 183., 184., 185., 186., 187., 188.,
189., 190., 191., 192., 193., 194., 195., 196., 197.,
198., 199., 200., 201., 202., 203., 204., 205., 206.,
207., 208., 209., 210., 211., 212., 213., 214., 215.,
216., 217., 218., 219., 220., 221., 222., 223., 224.,
225., 226., 227., 228., 229., 230., 231., 232., 233.,
234., 235., 236., 237., 238., 239., 240., 241., 242.,
243., 244., 245., 246., 247., 248., 249.])
```

Alternatively, you can use `genfromtxt()` and set the conversion of your header, by default, to `nan`. Then, you can detect how many lines are occupied by your header and use the `skip_header` parameter to skip them, as follows:

```
In [16]: b = np.genfromtxt("My_file.txt", delimiter='\t')
         b
Out[16]: array([ nan, nan, nan, nan, 0., 1., 2., 3., 4.,
         5., 6., 7., 8., 9., 10., 11., 12., 13.,
         14., 15., 16., 17., 18., 19., 20., 21., 22.,
         23., 24., 25., 26., 27., 28., 29., 30., 31.,
         32., 33., 34., 35., 36., 37., 38., 39., 40.,
         41., 42., 43., 44., 45., 46., 47., 48., 49.,
         50., 51., 52., 53., 54., 55., 56., 57., 58.,
         59., 60., 61., 62., 63., 64., 65., 66., 67.,
         68., 69., 70., 71., 72., 73., 74., 75., 76.,
         77., 78., 79., 80., 81., 82., 83., 84., 85.,
         86., 87., 88., 89., 90., 91., 92., 93., 94.,
         95., 96., 97., 98., 99., 100., 101., 102., 103.,
         104., 105., 106., 107., 108., 109., 110., 111., 112.,
         113., 114., 115., 116., 117., 118., 119., 120., 121.,
         122., 123., 124., 125., 126., 127., 128., 129., 130.,
         131., 132., 133., 134., 135., 136., 137., 138., 139.,
         140., 141., 142., 143., 144., 145., 146., 147., 148.,
         149., 150., 151., 152., 153., 154., 155., 156., 157.,
         158., 159., 160., 161., 162., 163., 164., 165., 166.,
         167., 168., 169., 170., 171., 172., 173., 174., 175.,
         176., 177., 178., 179., 180., 181., 182., 183., 184.,
         185., 186., 187., 188., 189., 190., 191., 192., 193.,
         194., 195., 196., 197., 198., 199., 200., 201., 202.,
         203., 204., 205., 206., 207., 208., 209., 210., 211.,
         212., 213., 214., 215., 216., 217., 218., 219., 220.,
         221., 222., 223., 224., 225., 226., 227., 228., 229.,
```

```
           230., 231., 232., 233., 234., 235., 236., 237., 238.,
           239., 240., 241., 242., 243., 244., 245., 246., 247.,
           248., 249.])
```

Similarly, you can use the `loadtxt()`, `genfromtxt()`, and `savetxt()` functions to load and save `.csv` files. The only thing you have to remember is to use a comma as a delimiter, as follows:

```
In [17]: data_csv = np.loadtxt("MyData.csv", delimiter=',')
In [18]: data_csv[1:3]
Out[18]: array([[ 0.21982, 0.31271, 0.66934, 0.06072, 0.77785, 0.59984,
                  0.82998, 0.77428, 0.73216, 0.29968],
                [ 0.78866, 0.61444, 0.0107 , 0.37351, 0.77391, 0.76958,
                  0.46845, 0.76387, 0.70592, 0.0851 ]])
In [19]: np.shape(data_csv)
Out[19]: (15, 10)
In [20]: np.savetxt('MyData1.csv',data_csv, delimiter = ',')
In [21]: data_csv1 = np.genfromtxt("MyData1.csv", delimiter = ',')
In [22]: data_csv1[1:3]
Out[22]: array([[ 0.21982, 0.31271, 0.66934, 0.06072, 0.77785, 0.59984,
                  0.82998, 0.77428, 0.73216, 0.29968],
                [ 0.78866, 0.61444, 0.0107 , 0.37351, 0.77391, 0.76958,
                  0.46845, 0.76387, 0.70592, 0.0851 ]])
In [23]: np.shape(data_csv1)
Out[23]: (15, 10)
```

When you load the `.txt` files, they return a `numpy` array with the values by default, as you can see in the following code:

```
In [24]: print (type(a))
         print (type(b))
         <class 'numpy.ndarray'>
         <class 'numpy.ndarray'>
```

You can change your data from an array to a list with the `tolist()` method and save it as a new file with `savetxt()` by using different delimiters, as follows:

```
In [25]: c = a.tolist()
         c
Out[25]: [0.0,
          1.0,
          2.0,
          3.0,
          4.0,
          5.0,
          6.0,
          7.0,
```

```
                8.0,
                9.0,
                10.0,
                11.0,
                12.0,
                13.0,
                14.0,
                15.0,
                16.0,
                17.0,
                18.0,
                ...
In [26]: np.savetxt('My_List.txt',c, delimiter=';')
In [27]: myList = np.loadtxt("My_List.txt", delimiter=';')
            type(myList)
Out[27]: numpy.ndarray
```

After you save the list in `My_List.txt`, you can load this file with `loadtxt()` and it will return as a `numpy` array once more. If you want to return a string representation of your array, you can use `array_str()`, `array_repr()`, or `array2string()` method, as follows:

```
In [28]: d = np.array_str(a,precision=1)
            d
Out[28]: '[ 0. 1. 2. 3. 4. 5. 6. 7. 8. 9. 10. 11.\n 12. 13. 14. 15. 16. 17.
18. 19. 20. 21. 22. 23.
            24. 25. 26. ... ]'
```

Even though `array_str()` and `array_repr()` look the same, `array_str()` returns a string representation of the data inside the array, while `array_repr()` actually returns the string representation of an array. Therefore, `array_repr()` returns information about the type of array and its data type. Both of these functions use `array2string()` internally, which is actually the most flexible version of the string formatting functions as it has more optional parameters. The following code block loads the Boston housing data set with `load_boston()` method:

```
In [29]: from sklearn.datasets import load_boston
            dataset = load_boston()
            dataset
```

In this chapter, you will practice exploratory data analysis on a sample dataset from the `sklearn.datasets` package. The dataset is about Boston house prices. In the preceding code, the `load_boston()` function is imported from the `sklearn.datasets` package, and as you can see, it returns a dictionary-like object with the attributes `DESCR`, `data`, `feature_names`, and `target`. The details of these attributes are as follows:

Attribute	Explanation
DESCR	Full description of the dataset
data	Feature columns
feature_names	Feature names
target	Label data

```
Out[29]:  {'data': array([[6.3200e-03, 1.8000e+01, 2.3100e+00, ..., 1.5300e+01, 3.9690e+02,
                 4.9800e+00],
                [2.7310e-02, 0.0000e+00, 7.0700e+00, ..., 1.7800e+01, 3.9690e+02,
                 9.1400e+00],
                [2.7290e-02, 0.0000e+00, 7.0700e+00, ..., 1.7800e+01, 3.9283e+02,
                 4.0300e+00],
                ...,
                [6.0760e-02, 0.0000e+00, 1.1930e+01, ..., 2.1000e+01, 3.9690e+02,
                 5.6400e+00],
                [1.0959e-01, 0.0000e+00, 1.1930e+01, ..., 2.1000e+01, 3.9345e+02,
                 6.4800e+00],
                [4.7410e-02, 0.0000e+00, 1.1930e+01, ..., 2.1000e+01, 3.9690e+02,
                 7.8800e+00]]),
          'target': array([24. , 21.6, 34.7, 33.4, 36.2, 28.7, 22.9, 27.1, 16.5, 18.9, 15. ,
                 18.9, 21.7, 20.4, 18.2, 19.9, 23.1, 17.5, 20.2, 18.2, 13.6, 19.6,
                 15.2, 14.5, 15.6, 13.9, 16.6, 14.8, 18.4, 21. , 12.7, 14.5, 13.2,
                 13.1, 13.5, 18.9, 20. , 21. , 24.7, 30.8, 34.9, 26.6, 25.3, 24.7,
                 21.2, 19.3, 20. , 16.6, 14.4, 19.4, 19.7, 20.5, 25. , 23.4, 18.9,
                 35.4, 24.7, 31.6, 23.3, 19.6, 18.7, 16. , 22.2, 25. , 33. , 23.5,
                 19.4, 22. , 17.4, 20.9, 24.2, 21.7, 22.8, 23.4, 24.1, 21.4, 20. ,
                 20.8, 21.2, 20.3, 28. , 23.9, 24.8, 22.9, 23.9, 26.6, 22.5, 22.2,
                 23.6, 28.7, 22.6, 22. , 22.9, 25. , 20.6, 28.4, 21.4, 38.7, 43.8,
                 33.2, 27.5, 26.5, 18.6, 19.3, 20.1, 19.5, 19.5, 20.4, 19.8, 19.4,
                 21.7, 22.8, 18.8, 18.7, 18.5, 18.3, 21.2, 19.2, 20.4, 19.3, 22. ,
                 20.3, 20.5, 17.3, 18.8, 21.4, 15.7, 16.2, 18. , 14.3, 19.2, 19.6,
                 23. , 18.4, 15.6, 18.1, 17.4, 17.1, 13.3, 17.8, 14. , 14.4, 13.4,
                 15.6, 11.8, 13.8, 15.6, 14.6, 17.8, 15.4, 21.5, 19.6, 15.3, 19.4,
                 17. , 15.6, 13.1, 41.3, 24.3, 23.3, 27. , 50. , 50. , 50. , 22.7,
                 25. , 50. , 23.8, 23.8, 22.3, 17.4, 19.1, 23.1, 23.6, 22.6, 29.4,
                 23.2, 24.6, 29.9, 37.2, 39.8, 36.2, 37.9, 32.5, 26.4, 29.6, 50. ,
                 32. , 29.8, 34.9, 37. , 30.5, 36.4, 31.1, 29.1, 50. , 33.3, 30.3,
                 34.6, 34.9, 32.9, 24.1, 42.3, 48.5, 50. , 22.6, 24.4, 22.5, 24.4,
                 20. , 21.7, 19.3, 22.4, 28.1, 23.7, 25. , 23.3, 28.7, 21.5, 23. ,
                 26.7, 21.7, 27.5, 30.1, 44.8, 50. , 37.6, 31.6, 46.7, 31.5, 24.3,
                 31.7, 41.7, 48.3, 29. , 24. , 25.1, 31.5, 23.7, 23.3, 22. , 20.1,
                 22.2, 23.7, 17.6, 18.5, 24.3, 20.5, 24.5, 26.2, 24.4, 24.8, 29.6,
                 42.8, 21.9, 20.9, 44. , 50. , 36. , 30.1, 33.8, 43.1, 48.8, 31. ,
                 36.5, 22.8, 30.7, 50. , 43.5, 20.7, 21.1, 25.2, 24.4, 35.2, 32.4,
                 32. , 33.2, 33.1, 29.1, 35.1, 45.4, 35.4, 46. , 50. , 32.2, 22. ,
```

In this section, you learned how to load and save files with `numpy` functions. In the next section, you will explore the Boston housing dataset.

Exploring our dataset

In this section, you will explore and perform quality checks on the dataset. You will check what your data shape is, as well as its data types, any missing/NaN values, how many feature columns you have, and what each column represents. Let's start by loading the data and exploring it:

```
In [30]: from sklearn.datasets import load_boston
         dataset = load_boston()
         samples,label, feature_names = dataset.data , dataset.target ,
dataset.feature_names
In [31]: samples.shape
Out[31]: (506, 13)
In [32]: label.shape
Out[32]: (506,)
In [33]: feature_names
Out[33]: array(['CRIM', 'ZN', 'INDUS', 'CHAS', 'NOX', 'RM', 'AGE', 'DIS',
'RAD',
                'TAX', 'PTRATIO', 'B', 'LSTAT'],
                dtype='<U7')
```

In the preceding code, you load the dataset and parse the attributes of your dataset. This shows us that we have 506 samples with 13 features and that we have 506 labels (regression targets). If you want to read the dataset's description, you can use `print(dataset.DESCR)`. As the output of this code is too long to put here, you can check out the various features and their descriptions in the following screenshot:

```
Data Set Characteristics:

    :Number of Instances: 506

    :Number of Attributes: 13 numeric/categorical predictive

    :Median Value (attribute 14) is usually the target

    :Attribute Information (in order):
        - CRIM      per capita crime rate by town
        - ZN        proportion of residential land zoned for lots over 25,000 sq.ft.
        - INDUS     proportion of non-retail business acres per town
        - CHAS      Charles River dummy variable (= 1 if tract bounds river; 0 otherwise)
        - NOX       nitric oxides concentration (parts per 10 million)
        - RM        average number of rooms per dwelling
        - AGE       proportion of owner-occupied units built prior to 1940
        - DIS       weighted distances to five Boston employment centres
        - RAD       index of accessibility to radial highways
        - TAX       full-value property-tax rate per $10,000
        - PTRATIO   pupil-teacher ratio by town
        - B         1000(Bk - 0.63)^2 where Bk is the proportion of blacks by town
        - LSTAT     % lower status of the population
        - MEDV      Median value of owner-occupied homes in $1000's
```

As shown in the first chapter, you can use `dtype` to check the data type of the array. As seen in the following code, we have a numeric `float64` data type in each column of the sample and a label. Checking the data type is a very important step—you may realize that there's some inconsistency between the types and the column description, or you may want to decrease the memory size of your array by changing its data type if you think that you can still achieve your goal with less precise values:

```
In [35]: print(samples.dtype)
         print(label.dtype)
         float64
         float64
```

Missing value handling is slightly different in Python packages. The `numpy` library doesn't have a missing value. If you have missing values in your dataset, they will convert to NaN after you import them. A very common method in NumPy is to use masked arrays in order to disregard NaN values, which was shown to you in the first chapter:

```
In [36]: np.isnan(samples)
Out[36]: array([[False, False, False, ..., False, False, False],
                [False, False, False, ..., False, False, False],
                [False, False, False, ..., False, False, False],
                ...,
```

```
                      [False, False, False, ..., False, False, False],
                      [False, False, False, ..., False, False, False],
                      [False, False, False, ..., False, False, False]],
        dtype=bool)
In [37]: np.isnan(np.sum(samples))
Out[37]: False
In [38]: np.isnan(np.sum(label))
Out[38]: False
```

To test element-wise for NaN values in your data, you can use the isnan() method. This method will return a Boolean array. It can be cumbersome for large arrays to detect whether it will return true or not. In such cases, you can use the np.sum() of your array as a parameter for isnan() so that it will return a single Boolean value for the result.

In this section, you explored data at a high level and performed a generic quality check. In the next section, we will continue with basic statistics.

Looking at basic statistics

In this section, you will start with the first step in statistical analysis by calculating the basic statistics of your dataset. Even though NumPy has limited built-in statistical functions, we can leverage its usage with SciPy. Before we start, let's describe how our analysis will flow. All of the feature columns and label columns are numerical, but you may have noticed that the **Charles River dummy variable (CHAS)** column has binary values *(0,1)*, which means that it's actually encoded from categorical data. When you analyze your dataset, you can separate your columns into Categorical and Numerical. In order to analyze them all together, one type should be converted to another. If you have a categorical value and you want to convert it into a numeric value, you can do so by converting each category to a numerical value. This process is called **encoding**. On the other hand, you can perform binning by transforming your numerical values into category counterparts, which you create by splitting your data into intervals.

We will start our analysis by exploring its features one by one. In statistics, this method is known as univariate analysis. The purpose of univariate analysis mainly centered around description. We will calculate the minimum, maximum, range, percentiles, mean, and variance, and then we will plot some histograms and analyze the distribution of each feature. We will touch upon the concept of skewness and kurtosis and then look at the importance of trimming. After finishing our univariate analysis, we will continue with bivariate analysis, which means simultaneously analyzing two features. To do this, we will explore the relationship between two sets of features:

```
In [39]: np.set_printoptions(suppress=True, linewidth=125)
```

```
        minimums = np.round(np.amin(samples, axis=0), decimals=1)
        maximums = np.round(np.amax(samples, axis=0), decimals=1)
        range_column = np.round(np.ptp(samples, axis=0), decimals=1)
        mean = np.round(np.mean(samples, axis=0), decimals=1)
        median = np.round(np.median(samples, axis=0), decimals=1)
        variance = np.round(np.var(samples, axis=0), decimals=1)
        tenth_percentile = np.round(np.percentile(samples, 10, axis=0),
decimals = 1)
        ninety_percentile = np.round(np.percentile(samples, 90 ,axis=0),
decimals = 1)
In [40]: range_column
Out[40]: array([ 89. , 100. ,   27.3,   1. ,   0.5,   5.2,  97.1,  11. ,
23. , 524. ,   9.4, 396.6,
36.2])
```

```
In [41]: Basic_Statistics = np.vstack((minimums,maximums,range_column,mean,median, variance, tenth_percentile,ninety_percentile))
         Basic_Statistics
Out[41]: array([[   0. ,     0. ,    0.5,     0. ,   0.4,    3.6,    2.9,    1.1,    1. ,   187. ,    12.6,     0.3,
            1.7],
         [  89. ,   100. ,   27.7,     1. ,   0.9,    8.8,  100. ,   12.1,   24. ,   711. ,    22. ,   396.9,
           38. ],
         [  89. ,   100. ,   27.3,     1. ,   0.5,    5.2,   97.1,   11. ,   23. ,   524. ,     9.4,  396.6,
           36.2],
         [   3.6,    11.4,   11.1,    0.1,   0.6,    6.3,   68.6,    3.8,    9.5,   408.2,    18.5,  356.7,
           12.7],
         [   0.3,     0. ,    9.7,     0. ,   0.5,    6.2,   77.5,    3.2,    5. ,   330. ,    19. ,   391.4,
           11.4],
         [  73.8,   542.9,   47. ,    0.1,   0. ,    0.5,  790.8,    4.4,   75.7, 28348.6,     4.7, 8318.3,
           50.9],
         [   0. ,     0. ,    2.9,     0. ,   0.4,    5.6,   27. ,    1.6,    3. ,   233. ,    14.8,  290.3,
            4.7],
         [  10.5,    42.5,   19.6,     0. ,   0.7,    7.2,   98.8,    6.8,   24. ,   666. ,    20.9,  396.9,
           23. ]])
```

In the preceding code, we begin by setting our print options with the set_printoptions() method in order to see rounded decimals and have a line width big enough to fit all of the columns. To calculate basic statistics, we use numpy functions, such as amin(), amax() , mean(), median(), var() , percentile(), and ptp(). All of the calculations are done column-wise as each column represents a feature. The final array looks a bit sloppy and uninformative as you can't see which row shows which statistics:

```
In [42]: stat_labels =  'minm', 'maxm', 'rang', 'mean','medi', 'vari','50%t','90%t'

In [43]: print("          F1     F2     F3     F4     F5     F6     F7     F8     F9     F10    F11    F12   F13  ")
         for stat_labels , row in zip(stat_labels,Basic_Statistics):
             print('%s [%s]' % (stat_labels, ''.join('%07s' % i for i in row)))

                 F1     F2     F3     F4     F5     F6     F7     F8     F9     F10    F11    F12   F13
         minm [   0.0    0.0    0.5    0.0    0.4    3.6    2.9    1.1    1.0  187.0   12.6    0.3    1.7]
         maxm [  89.0  100.0   27.7    1.0    0.9    8.8  100.0   12.1   24.0  711.0   22.0  396.9   38.0]
         rang [  89.0  100.0   27.3    1.0    0.5    5.2   97.1   11.0   23.0  524.0    9.4  396.6   36.2]
         mean [   3.6   11.4   11.1    0.1    0.6    6.3   68.6    3.8    9.5  408.2   18.5  356.7   12.7]
         medi [   0.3    0.0    9.7    0.0    0.5    6.2   77.5    3.2    5.0  330.0   19.0  391.4   11.4]
         vari [  73.8  542.9   47.0    0.1    0.0    0.5  790.8    4.4   75.728348.6    4.7 8318.3   50.9]
         50%t [   0.0    0.0    2.9    0.0    0.4    5.6   27.0    1.6    3.0  233.0   14.8  290.3    4.7]
         90%t [  10.5   42.5   19.6    0.0    0.7    7.2   98.8    6.8   24.0  666.0   20.9  396.9   23.0]
```

In order to print an aligned numpy array, you can use the `zip()` function to add your row labels and print the column labels before your array. In SciPy, you can use many more statistical functions to calculate basic statistics. SciPy supplies the `describe()` function, which returns several descriptive statistics of the given array. You can calculate nobs, minimum, maximum, mean, variance, skewness, and kurtosis with a single function and parse them separately, as follows:

```
In [44]: from scipy import stats
         arr= stats.describe(samples, axis=0)
         arr

Out[44]: DescribeResult(nobs=506, minmax=(array([  0.00632,   0.      ,   0.46   ,   0.      ,   0.385  ,   3.561  ,   2.9    ,   1.1296 ,
         1.      , 187.      ,
              12.6    ,   0.32   ,   1.73   ]), array([ 88.9762, 100.      ,  27.74   ,   1.      ,   0.871  ,   8.78   , 100.      ,  12.126  ,
         5.  24.      , 711.      ,  22.      ,
             396.9    ,  37.97   ])), mean=array([  3.59376071,  11.36363636,  11.13677866,   0.06916996,   0.55469506,   6.28463439,
         68.57490119,   3.79504269,
               9.54940711, 408.23715415,  18.4555336 , 356.67403162,  12.65306324]), variance=array([  73.90467096,  543.93681368,
         47.06444247,    0.06451297,    0.01342764,    0.49367085,  792.35839851,
                4.43401514,   75.81636598, 28404.75948812,    4.68698912, 8334.75226292,   50.99475951]), skewness=array([  5.222
         03907,   2.21906306,   0.29414628,   3.39579929,   0.72714416,   0.40241467,  -0.59718559,   1.00877876,   1.00183349,
                0.66796827,  -0.79994453,  -2.88179835,   0.90377074]), kurtosis=array([36.88811011,   3.97994877,  -1.23321847,   9.5314528
         4, -0.07586422,   1.86102697,  -0.97001393,   0.47129857,  -0.8705205 ,
                -1.14298488,  -0.29411638,   7.14376929,   0.47654476]))
```

The following code block calculates the basic statistics separately and stacks them into a final array:

```
In [45]: minimum = arr.minmax[0]
         maximum = arr.minmax[1]
         mean = arr.mean
         median = np.round(np.median(samples,axis = 0), decimals = 1)
         variance = arr.variance
         tenth_percentile = stats.scoreatpercentile(samples, per = 10, axis
= 0)

         ninety_percentile = stats.scoreatpercentile(samples, per =90, axis
= 0)

         rng = stats.iqr(samples, rng = (20,80), axis = 0)
         np.set_printoptions(suppress = True, linewidth = 125)
         Basic_Statistics1 = np.round(np.vstack((minimum,maximum,rng,
mean,median,variance,tenth_percentile,ninety_percentile)),
Basic_Statistics1.shape
Out[45]: (8, 13)
In [46]: stat_labels1 = ['minm', 'maxm', 'rang', 'mean', 'medi', 'vari',
'50%t', '90%t']
```

```
In [47]: np.set_printoptions(suppress= True, linewidth= 125)
         print("             F1     F2     F3     F4     F5     F6     F7     F8     F9    F10    F11     F12    F13")
         for stat_labels1, row1 in zip(stat_labels1, Basic_Statistics1):
             print ('%s [%s]' % (stat_labels1, ''.join('%07s' % a for a in row1)))

                  F1     F2     F3     F4     F5     F6     F7     F8     F9     F10    F11     F12    F13
         minm [   0.0    0.0    0.5    0.0    0.4    3.6    2.9    1.1    1.0   187.0   12.6    0.3    1.7]
         maxm [  89.0  100.0   27.7    1.0    0.9    8.8  100.0   12.1   24.0   711.0   22.0  396.9   38.0]
         rang [   5.4   20.0   13.7    0.0    0.2    0.9   57.8    3.7   20.0   393.0    3.6   32.6   11.8]
         mean [   3.6   11.4   11.1    0.1    0.6    6.3   68.6    3.8    9.5   408.2   18.5  356.7   12.7]
         medi [   0.3    0.0    9.7    0.0    0.5    6.2   77.5    3.2    5.0   330.0   19.0  391.4   11.4]
         vari [  73.9  543.9   47.1    0.1    0.0    0.5  792.4    4.4  75.8 28404.8    4.7 8334.8   51.0]
         50%t [   0.0    0.0    2.9    0.0    0.4    5.6   27.0    1.6    3.0   233.0   14.8  290.3    4.7]
         90%t [  10.5   42.5   19.6    0.0    0.7    7.2   98.8    6.8   24.0   666.0   20.9  396.9   23.0]
```

In contrast to NumPy, we can use the `iqr()` function to calculate the range. This function calculates the interquartile range of the data along the specified axis and **range** (rng parameter). By default, *rng = (25, 75)*, which means that the function will calculate the difference between the 75th and 25th percentile values of the data. In order to return the same output as the `numpy.ptp()` function, you can use rng =(0, 100), as this will calculate the range of all given data. We used `stat.scoreatpercentile()` as an equivalent to the `numpy.percentile()` method in order to calculate the 10th and 90th percentile values of the features. If you take a glance at the results, you will notice that we have very high variance in almost every feature. You can see that the range values decreases considerably as we limit the range calculation by passing the parameters as `(20,80)`. This also shows you that we have many extreme values on both sides of the distributed features. We can conclude from our results that, for most of the features, the mean is higher than the median, which shows us that the distribution of these features is skewed to the right. In the next section, you will see this clearly when we plot the histograms and then deeply analyze the skewness and kurtosis of these features.

Computing histograms

A histogram is a visual representation of the distribution of numerical data. Karl Pearson first introduced this concept more than a century ago. A histogram is a kind of bar chart that is used for continuous data, while a bar chart visually represents categorical variables. As a first step, you need to divide your entire range of values into a series of intervals (bins). Each bin has to be adjacent and none of them can overlap. In general, bin sizes are equal, and the rule of thumb for the number of bins to include is 5–20. This means that if you have more than 20 bins, your graph will be hard to read. On the contrary, if you have fewer than 5 bins, then your graph will give very little insight into the distribution of your data:

```
In [48]: %matplotlib notebook
```

```
%matplotlib notebook
import matplotlib.pyplot as plt
NOX = samples[:,5:6]
plt.hist(NOX,bins ='auto')
plt.title("Distribution nitric oxides concentration (parts per 10
million)")
plt.show()
```

In preceding code, we plot the histogram for the feature NOX. Bins calculations done automatically as follows:

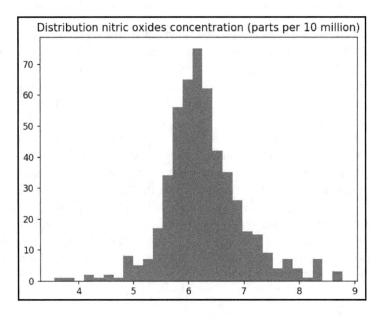

You can plot your histogram with `pyplot.hist()` by giving it your sliced array as the first argument. The *y*-axis shows the count of how many values fall into each interval (bin) in your dataset and the *x*-axis represents their values. By setting `normed=True`, you can see the percentage of your bins as follows:

```
In [49]: plt.hist(NOX,bins ='auto', normed = True)
         plt.title("Distribution nitric oxides concentration (parts per 10
million)")
         plt.show()
```

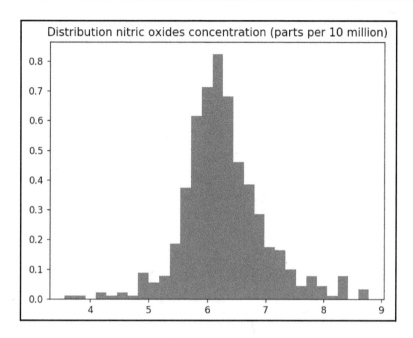

When you look at the histogram, it can be hard to calculate the size of each bin and its edges. As `pyplot.hist()` returns a tuple that includes this information, you can easily get these values, as follows:

```
In [50]: import matplotlib.pyplot as plt
         NOX = samples[:,5:6]
         n, bins, patches = plt.hist(NOX, bins='auto')
         print('Bin Sizes')
         print(n)
         print('Bin Edges')
         print(bins)
```

In the preceding code, we print how many numbers included in each bin and what are the edges of bins as follows:

```
Bin Sizes
[ 1.  1.  0.  2.  1.  2.  1.  8.  5.  7. 17. 34. 56. 65. 75. 62. 42. 35. 26. 16. 15.  9.  4.  7.  4.  1.  7.  0.  3.]
Bin Edges
[3.561      3.74096552 3.92093103 4.10089655 4.28086207 4.46082759 4.6407931  4.82075862 5.00072414 5.18068966 5.36065517
 5.54062069 5.72058621 5.90055172 6.08051724 6.26048276 6.44044828 6.62041379 6.80037931 6.98034483 7.16031034 7.34027586
 7.52024138 7.7002069  7.88017241 8.06013793 8.24010345 8.42006897 8.60003448 8.78      ]
```

Let's interpret the preceding output. The first bin's size is 1, which means that there is only a single value drop in this bin. The interval of the first bin is between 3.561 and 3.74096552. In order to make them tidier, you can use the following code block, which stacks these two arrays meaningfully (bin_interval, bin_size):

```
In [51]: bins_string = bins.astype(np.str)
         n_string = n.astype(np.str)
         lists = []
         for i in range(0, len(bins_string)-1):
             c = bins_string[i]+ "-" + bins_string[i+1]
             lists.append(c)
         new_bins = np.asarray(lists)
         Stacked_Bins = np.vstack((new_bins, n_string)).T
         Stacked_Bins
```

```
Out[51]: array([['3.561-3.740965517241379', '1.0'],
               ['3.740965517241379-3.9209310344827584', '1.0'],
               ['3.9209310344827584-4.100896551724138', '0.0'],
               ['4.100896551724138-4.280862068965517', '2.0'],
               ['4.280862068965517-4.4608275862068965', '1.0'],
               ['4.4608275862068965-4.640793103448276', '2.0'],
               ['4.640793103448276-4.820758620689655', '1.0'],
               ['4.820758620689655-5.0007241379310345', '8.0'],
               ['5.0007241379310345-5.180689655172413', '5.0'],
               ['5.180689655172413-5.360655172413793', '7.0'],
               ['5.360655172413793-5.540620689655173', '17.0'],
               ['5.540620689655173-5.720586206896551', '34.0'],
               ['5.720586206896551-5.90055172413793', '56.0'],
               ['5.90055172413793-6.08051724137931', '65.0'],
               ['6.08051724137931-6.2604827586206895', '75.0'],
               ['6.2604827586206895-6.440448275862069', '62.0'],
               ['6.440448275862069-6.620413793103448', '42.0'],
               ['6.620413793103448-6.800379310344827', '35.0'],
               ['6.800379310344827-6.980344827586206', '26.0'],
               ['6.980344827586206-7.160310344827586', '16.0'],
               ['7.160310344827586-7.340275862068966', '15.0'],
               ['7.340275862068966-7.520241379310344', '9.0'],
               ['7.520241379310344-7.700206896551724', '4.0'],
               ['7.700206896551724-7.880172413793103', '7.0'],
               ['7.880172413793103-8.060137931034483', '4.0'],
               ['8.060137931034483-8.24010344827586', '1.0'],
               ['8.24010344827586-8.420068965517242', '7.0'],
               ['8.420068965517242-8.60003448275862', '0.0'],
               ['8.60003448275862-8.78', '3.0']], dtype='<U36')
```

Deciding on the number of bins and their width is very important. Some theoreticians experiment with this to determine the best fit. The following table shows you the most popular estimators. You can set the estimator to the `bins` parameter in `numpy.histogram()` to change the bin calculation accordingly. These methods are implicitly supported by the `pyplot.hist()` function, as its arguments are passed to `numpy.histogram()`:

Estimator	Formula
Freedman–Diaconis estimator	$h = 2\frac{IQR(x)}{n^{1/3}}$
Doane's formula	$k = 1 + log_2(n) + log_2\left(1 + \frac{|g_1|}{\sigma_{g1}}\right)$
Rice rule	$k = 2n^{1/3}$
Scott's normal reference rule	$h = \frac{3.5(\sigma)}{n^{1/3}}$
Sturges' formula	$k = log_2 n + 1$
Square-root choice	$k = \sqrt{n}$

IQR = Interquartile range

g_1 = *Estimated third-moment skewness of the distribution*

All of these methods have different strengths. For example, Sturges' formula is optimal for Gaussian data. Rice's rule is a simplified version of Sturges's formula and assumes an approximately normal distribution, so it may not perform well if the data is not normally distributed. Doane's formula is an improved version of Sturges's formula, especially with nonnormal distributions. The Freedman–Diaconis Estimator is the modified version of Scott's rule, where he replaced the 3.5 standard deviations with 2 *IQR*, which makes the formula less sensitive to the outliers. The square-root choice is a very common method, which is used by many tools for its speed and simplicity. In `numpy.histogram()`, there is another option called `'auto'`, which gives us the maximum of the Sturges and Freedman–Diaconis estimator methods. Let's see how these methods will affect our histogram:

```
In [52]: fig, ((ax1, ax2, ax3),(ax4,ax5,ax6)) =
plt.subplots(2,3,sharex=True)
        axs = [ax1,ax2,ax3,ax4,ax5,ax6]
        list_methods = ['fd','doane', 'scott', 'rice', 'sturges','sqrt']
        plt.tight_layout(pad=1.1, w_pad=0.8, h_pad=1.0)
        for n in range(0, len(axs)):
            axs[n].hist(NOX,bins = list_methods[n])
            axs[n].set_title('{}'.format(list_methods[n]))
```

In the preceding code, we compile six histograms and all of them share the same *x* axis as follows:

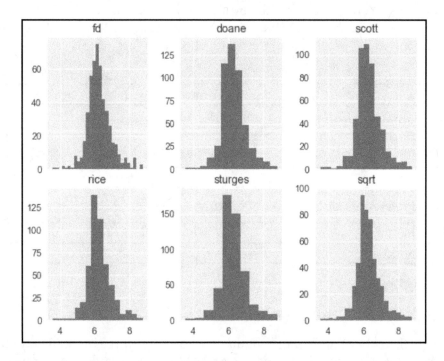

All of the histograms represent the same feature but with different binning methods. For example, the fd method shows that the data looks close to its normal distribution, while on the other hand, the doane method looks more like the student's *t*-distribution. In addition, the sturges method creates two little bins, so it's very hard to analyze the data. When we were looking at the basic statistics, we stated that most of the features had a mean that is higher than their median, so they skewed their data to the right. However, if you look at the sturges, rice, and sqrt methods, it's very hard to say that. Nevertheless, this is obvious when we plot the histogram with auto bins:

```
In [53]: import numpy as np
         samples_new = np.delete(samples, 3 , axis=1)
         samples_new.shape
Out[53]: (506, 12)
In [54]: %matplotlib notebook
         %matplotlib notebook
         import matplotlib.pyplot as plt
         fig,((ax1, ax2 , ax3),(ax4, ax5, ax6), (ax7, ax8, ax9), (ax10,
ax11, ax12)) = plt.subplots(4,3,                                 figsize =
(10,15))
```

```
axs =[ax1, ax2 , ax3, ax4, ax5, ax6, ax7, ax8, ax9, ax10, ax11,
ax12]

feature_names_new = np.delete(feature_names,3)
for n in range(0, len(axs)):
    axs[n].hist(samples_new[:,n:n+1], bins ='auto', normed = True)
    axs[n].set_title('{}'.format(feature_names[n]))
```

In the preceding code, we compile all the feature histograms in a single layout. This will help us to compare them easily. The output of the preceding code is as follows:

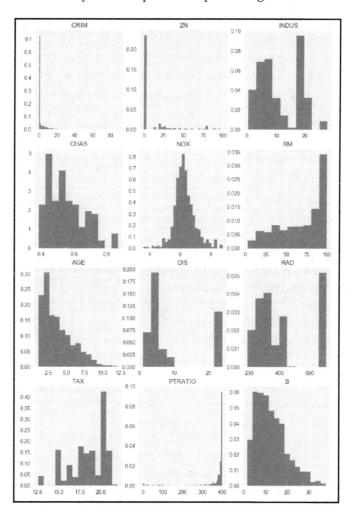

In the preceding code, we delete **CHAS** as it contains a binary value, which means that the histogram will not help us gain more insight about this feature. We also took out the feature's name from the feature list in order to plot the rest of the features correctly.

We can conclude from these graphs that, per capita, crime is very low in most towns, but that there are towns where this ratio is extremely high. Generally, residential land zones are lower than 25,000 ft. In many cases, non-retail business acres cover less than 10% of the total town. On the other hand, we can see that many towns have around 20% non-retail business acres. Nitric oxide concentration is very right-skewed, although there are some outliers here, which are very far from the mean. The average number of rooms per dwelling is between 5–7 rooms. Of the buildings that were built before 1940, more than 50% of them were occupied by their owners. Most of them don't live very far away from Boston's employment centers. More than 10% don't have good accessibility to radial highways. Property taxes are very high for a considerable number of people. Generally, classroom sizes are between 15–20 children. Most of the towns have a very similar proportion of black people living in them. Most of the townspeople have a lower economic status. You can draw many other conclusions by looking at these histograms. As you can see, histograms are very useful when you are describing how your data is distributed, and you can see the average, variance, and outliers. In the next section, we will focus on skewness and kurtosis.

Explaining skewness and kurtosis

In statistical analysis, a moment is a quantitative measure that describes the expected distance from a reference point. If the reference point is expected, then it's called a central moment. In statistics, the central moments are the moments that are related with the mean. The first and second moments are the mean and the variance, respectively. The mean is the average of your data points. The variance is the total deviation of each data point from the mean. In other words, the variance shows how your data is dispersed from the mean. The third central moment is skewness, which measures the asymmetry of the distribution of the mean. In standard normal distribution, skewness equals zero as it's symmetrical. On the other hand, if mean < median < mode, then there is negative skew, or left skew; likewise, if mode < median < mean, there is positive skew or right skew. You may be confused when you hear the terms right skewed or left skewed, and start imagining a distribution that leans to the right or left side. Technically, it is the other way around, so when you are talking about left skew, you should think about the tails (outliers) of the distribution on the left side, since your distribution actually looks like it's leaning to the right:

```
In [55]: %matplotlib notebook
         %matplotlib notebook
         from scipy.stats import skewnorm
```

```
fig, (ax1, ax2, ax3) = plt.subplots(1,3 ,figsize=(10,2.5))
x1 = np.linspace(skewnorm.ppf(0.01,-3), skewnorm.ppf(0.99,-3),100)
x2 = np.linspace(skewnorm.ppf(0.01,0), skewnorm.ppf(0.99,0),100)
x3 = np.linspace(skewnorm.ppf(0.01,3), skewnorm.ppf(0.99,3),100)
ax1.plot(skewnorm(-3).pdf(x1),'k-', lw=4)
ax2.plot(skewnorm(0).pdf(x2),'k-', lw=4)
ax3.plot(skewnorm(3).pdf(x3),'k-', lw=4)
ax1.set_title('Left Skew')
ax2.set_title('Normal Dist')
ax3.set_title('Right Skew')
```

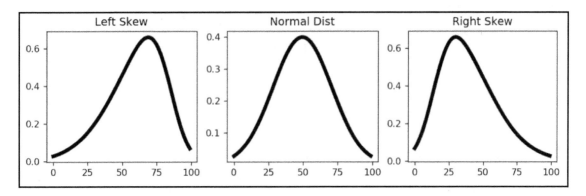

You can easily create a normal distribution with skewness by using the `skewnorm()` function. As you can see in the preceding code, we generate a percent point function (an inverse of the cumulative distribution function percentiles) with 100 values and then create lines with different skewness levels. You cannot directly conclude that the left skew is better than right skew or vice versa. When you analyze skewness of your data, you need to consider what these tails can cause. As an example, if you are analyzing a time series of stock returns, and after plotting it you see a right skew, you don't need to be concerned about getting higher returns than you expect since these outliers will not cause any risk to your trading strategy. A similar example can be when you're analyzing your server's response time. When you plot the probability density function of your response time, you aren't really interested in left tails as they represent quick response times.

The fourth central moment concerning the mean is kurtosis. Kurtosis describes the tails (outliers) in relation to how flat or thin the curve of the distribution is. In a normal distribution, kurtosis equals 3. There are three main types of kurtosis: Leptokurtic (thin), Mesokurtic, and Platykurtic (flat):

```
In [56]: %matplotlib notebook
         %matplotlib notebook
         import scipy
         from scipy import stats
```

```python
import matplotlib.pyplot as plt
fig, (ax1, ax2, ax3) = plt.subplots(1, 3 , figsize=(10,2))
axs= [ax1, ax2, ax3]
Titles = ['Mesokurtic', 'Lebtokurtic', 'Platykurtic']
#Mesokurtic Distribution - Normal Distribution
dist = scipy.stats.norm(loc=100, scale=5)
sample_norm = dist.rvs(size = 10000)
#leptokurtic Distribution
dist2 = scipy.stats.laplace(loc= 100, scale= 5)
sample_laplace = dist2.rvs(size= 10000)
#platykurtic Distribution
dist3 = scipy.stats.cosine(loc= 100, scale= 5)
sample_cosine = dist3.rvs(size= 10000)
samples = [sample_norm, sample_laplace, sample_cosine]

for n in range(0, len(axs)):
    axs[n].hist(samples[n],bins= 'auto', normed= True)
    axs[n].set_title('{}'.format(Titles[n]))
    print ("kurtosis of" + Titles[n])
    print(scipy .stats.describe(samples[n])[5])
```

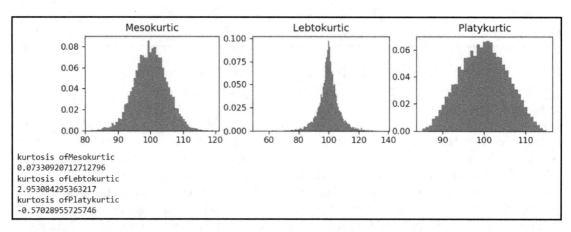

In the preceding code, you can see the differences in the shape of the distribution more clearly. All of the kurtosis values are normalized with the `stats.describe()` method, so the actual kurtosis value of the normal distribution is around 3, whereas 0.02 is the normalized version. Let's quickly check what the skewness and kurtosis values for our features are:

```python
In [57]: samples,label, feature_names = dataset.data , dataset.target ,
dataset.feature_names
         for n in range(0, len(feature_names_new)):
             kurt = scipy.stats.describe(samples[n])[5]
```

```
        skew = scipy.stats.describe(samples[n])[4]
        print (feature_names_new[n] + "-Kurtosis: {} Skewness: {}"
.format(kurt, skew))
```

```
    CRIM-Kurtosis: 2.102090573040533 Skewness: 1.9534138515494224
    ZN-Kurtosis: 2.8706349006925134 Skewness: 2.0753333576721893
    INDUS-Kurtosis: 2.9386308786131767 Skewness: 2.1061627843164086
    NOX-Kurtosis: 3.47131446484547 Skewness: 2.2172838215060517
    RM-Kurtosis: 3.461596258869246 Skewness: 2.2086627738768234
    AGE-Kurtosis: 3.395079726813977 Skewness: 2.1917520072643533
    DIS-Kurtosis: 1.9313625761956317 Skewness: 1.924572804475305
    RAD-Kurtosis: 1.7633603556547106 Skewness: 1.8601991629604233
    TAX-Kurtosis: 1.637076772210217 Skewness: 1.8266096199819994
    PTRATIO-Kurtosis: 1.7459544645159752 Skewness: 1.8679592455694167
    B-Kurtosis: 1.7375702020429316 Skewness: 1.8566444885400044
    LSTAT-Kurtosis: 1.8522036606250456 Skewness: 1.892802610207445
```

As we can see from the results, all of the features have a positive skewness, which indicates that they are skewed to the right. In respect of kurtosis, they all have positive values, especially NOX and RM, which have very high kurtoses. We can conclude that they all have a leptokurtic shape type, which indicates that the data is more concentrated on the mean. In the next section, we will focus on data trimming and calculate statistics with trimmed data.

Trimmed statistics

As you will have noticed in the previous section, the distributions of our features are very dispersed. Handling the outliers in your model is a very important part of your analysis. It is also very crucial when you look at descriptive statistics. You can be easily confused and misinterpret the distribution because of these extreme values. SciPy has very extensive statistical functions for calculating your descriptive statistics in regards to trimming your data. The main idea of using the trimmed statistics is to remove the outliers (tails) in order to reduce their effect on statistical calculations. Let's see how we can use these functions and how they will affect our feature distribution:

```
In [58]: np.set_printoptions(suppress= True, linewidth= 125)
        samples = dataset.data
        CRIM = samples[:,0:1]
        minimum = np.round(np.amin(CRIM), decimals=1)
        maximum = np.round(np.amax(CRIM), decimals=1)
        variance = np.round(np.var(CRIM), decimals=1)
        mean = np.round(np.mean(CRIM), decimals=1)
        Before_Trim = np.vstack((minimum, maximum, variance, mean))
        minimum_trim = stats.tmin(CRIM, 1)
```

```
        maximum_trim = stats.tmax(CRIM, 40)
        variance_trim = stats.tvar(CRIM, (1,40))
        mean_trim = stats.tmean(CRIM, (1,40))
        After_Trim =
np.round(np.vstack((minimum_trim,maximum_trim,variance_trim,mean_trim)),
decimals=1)
        stat_labels1 = ['minm', 'maxm', 'vari', 'mean']
        Basic_Statistics1 = np.hstack((Before_Trim, After_Trim))
        print ("     Before     After")
        for stat_labels1, row1 in zip(stat_labels1, Basic_Statistics1):
            print ('%s [%s]' % (stat_labels1, ''.join('%07s' % a for a in
row1)))
```

```
            Before    After
    minm [  0.0       1.0]
    maxm [ 89.0      38.4]
    vari [ 73.8      48.1]
    mean [  3.6       8.3]
```

To calculate the trimmed statistics, we used `tmin()`, `tmax()`, `tvar()`, and `tmean()`, as shown in the preceding code. All of these functions take limit values as a second parameter. In the `CRIM` feature, we can see many tails on the right side, so we limit the data to `(1, 40)` and then calculate the statistics. You can observe the difference by comparing the calculated statistics both before and after we have trimmed the values. As an alternative for `tmean()`, the `trim_mean()` function can be used if you want to slice your data proportionally from both sides. You can see how our mean and variance significantly changes after trimming the data. The variance is significantly decreased as we removed many extreme outliers between 40 and 89. The same trimming has a different effect on the mean, where the mean afterwards is more than doubled. The reason for this is that there were many zeros in the previous distribution, and by limiting the calculation between the values of 1 and 40, we removed all of these zeros, which resulted in a higher mean. Be advised that all of the preceding functions just trim your data on the fly while calculating these values, so the `CRIM` array is not trimmed. If you want to trim your data from both sides, you can use `trimboth()` and `trim1()` for one side. In both methods, instead of using limit values, you should use proportions as parameters. As an example, if you pass `proportiontocut` =0.2, it will slice your leftmost and rightmost values by 20%:

```
In [59]: %matplotlib notebook
         %matplotlib notebook
         import matplotlib.pyplot as plt
         CRIM_TRIMMED = stats.trimboth(CRIM, 0.2)
         fig, (ax1, ax2) = plt.subplots(1,2 , figsize =(10,2))
         axs = [ax1, ax2]
         df = [CRIM, CRIM_TRIMMED]
         list_methods = ['Before Trim', 'After Trim']
```

```
for n in range(0, len(axs)):
    axs[n].hist(df[n], bins = 'auto')
    axs[n].set_title('{}'.format(list_methods[n]))
```

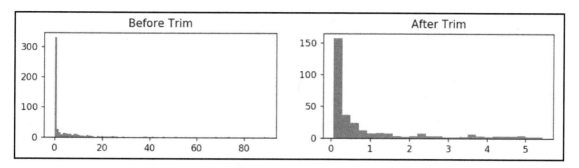

After trimming 20% of the values from both sides, we can interpret the distribution better and actually see that most of the values are between 0 and 1.5. It's hard to get this insight just by looking at the left histogram, as extreme values are dominating the histogram, and we can only see a single line around 0. As a result, `trimmed` functions are very useful in exploratory data analysis. In the next section, we will focus on box plots, which are very useful and popular graphical visuals for the descriptive analysis of data and outlier detection.

Box plots

Another important visual in exploratory data analysis is the box plot, also known as the box-and-whisker plot. It's built based on the five-number summary, which is the minimum, first quartile, median, third quartile, and maximum values. In a standard box plot, these values are represented as follows:

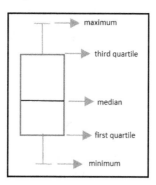

It's a very convenient way of comparing several distributions. In general, the whiskers of the plot generally extend to the extreme points. Alternatively, you can cut them with the 1.5 interquartile range. Let's check our CRIM and RM features:

```
In [60]: %matplotlib notebook
         %matplotlib notebook
         import matplotlib.pyplot as plt
         from scipy import stats
         samples = dataset.data
         fig, (ax1,ax2) = plt.subplots(1,2, figsize =(8,3))
         axs = [ax1, ax2]
         list_features = ['CRIM', 'RM']
         ax1.boxplot(stats.trimboth(samples[:,0:1],0.2))
         ax1.set_title('{}'.format(list_features[0]))
         ax2.boxplot(stats.trimboth(samples[:,5:6],0.2))
         ax2.set_title('{}'.format(list_features[1]))
```

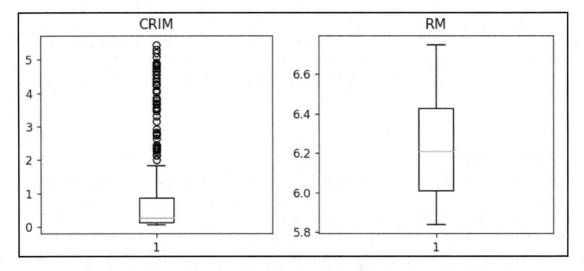

As you can see in the preceding code, the RM values have been continuously dispersed along the minimum and maximum so that the whiskers look like a line. You can easily detect that the median is almost in the middle between the first and third quartile. In CRIM, the extreme values do not continue along the minimum; they are more like individual outlying data points, so the representation is like a circle. You can see that these outliers are mostly after the third quartile and that the median is pretty close to the first quartile. From this, you can also conclude that the distribution is right-skewed. As a result, the box plot is a very useful alternative to the histogram, as you can easily analyze your distributions and compare several at once. In the next section, we will continue with bivariate analysis, and we will look at the correlation of label data with features.

Computing correlations

This section is dedicated to bivariate analysis, where you analyze two columns. In such cases, we generally investigate the association between these two variables, which is called **correlation**. Correlation shows the relationship between two variables and answers questions such as what will happen to variable A if variable B increases by 10%? In this section, we will explain how to calculate the correlation of our data and represent it in a two-dimensional scatter plot.

In general, correlation refers to any statistical dependency. A correlation coefficient is a quantitative value that calculates the measure of correlation. You can think of the relationship between correlation and a correlation coefficient as being of a similar relationship between a hygrometer and humidity. One of the most popular types of correlation coefficient is the Pearson product-moment correlation coefficient. The value of the correlation coefficient is between +1 and -1, where -1 indicates a strong negative linear relationship between two variables and +1 indicates a strong positive linear relationship between two variables. In NumPy, we can calculate the coefficient correlation matrix by using the `corrcoef()` method, as follows:

```
In [61]: np.set_printoptions(suppress= True, linewidth = 125)
         CorrelationCoef_Matrix = np.round(np.corrcoef(samples, rowvar=
False), decimals= 1)
         CorrelationCoef_Matrix
```

```
Out[61]: array([[ 1. , -0.2,  0.4, -0.1,  0.4, -0.2,  0.4, -0.4,  0.6,  0.6,  0.3, -0.4,  0.5],
               [-0.2,  1. , -0.5, -0. , -0.5,  0.3, -0.6,  0.7, -0.3, -0.3, -0.4,  0.2, -0.4],
               [ 0.4, -0.5,  1. ,  0.1,  0.8, -0.4,  0.6, -0.7,  0.6,  0.7,  0.4, -0.4,  0.6],
               [-0.1, -0. ,  0.1,  1. ,  0.1,  0.1,  0.1, -0.1, -0. , -0. , -0.1,  0. , -0.1],
               [ 0.4, -0.5,  0.8,  0.1,  1. , -0.3,  0.7, -0.8,  0.6,  0.7,  0.2, -0.4,  0.6],
               [-0.2,  0.3, -0.4,  0.1, -0.3,  1. , -0.2,  0.2, -0.2, -0.3, -0.4,  0.1, -0.6],
               [ 0.4, -0.6,  0.6,  0.1,  0.7, -0.2,  1. , -0.7,  0.5,  0.5,  0.3, -0.3,  0.6],
               [-0.4,  0.7, -0.7, -0.1, -0.8,  0.2, -0.7,  1. , -0.5, -0.5, -0.2,  0.3, -0.5],
               [ 0.6, -0.3,  0.6, -0. ,  0.6, -0.2,  0.5, -0.5,  1. ,  0.9,  0.5, -0.4,  0.5],
               [ 0.6, -0.3,  0.7, -0. ,  0.7, -0.3,  0.5, -0.5,  0.9,  1. ,  0.5, -0.4,  0.5],
               [ 0.3, -0.4,  0.4, -0.1,  0.2, -0.4,  0.3, -0.2,  0.5,  0.5,  1. , -0.2,  0.4],
               [-0.4,  0.2, -0.4,  0. , -0.4,  0.1, -0.3,  0.3, -0.4, -0.4, -0.2,  1. , -0.4],
               [ 0.5, -0.4,  0.6, -0.1,  0.6, -0.6,  0.6, -0.5,  0.5,  0.5,  0.4, -0.4,  1. ]])
```

Seaborn is a statistical data visualization library based on matplotlib which you can use to create very attractive and aesthetic statistical graphics. It`s very popular library with beautiful visualization with a perfect compatibility with popular packages, especially with pandas. You can use the heat map from the `seaborn` package to visualize the correlation coefficient matrix. It's very useful for detecting high-correlation coefficients when you have hundreds of features:

```
In [62]: CorrelationCoef_Matrix1 = np.round(np.corrcoef(samples, rowvar=
False), decimals= 1)
         CorrelationCoef_Matrix1
         import seaborn as sns; sns.set()
         ax = sns.heatmap(CorrelationCoef_Matrix1, cmap= "YlGnBu")
```

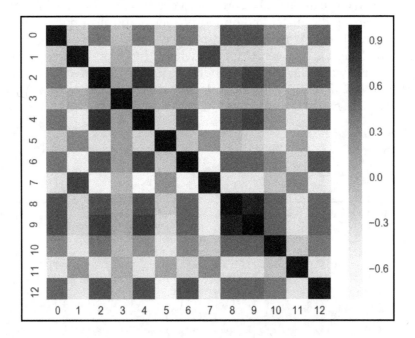

Here, we have the correlation coefficients of features of the label column. You can add an additional set of variables in the `corrcoef()` function as a second argument, as we did for the label column, which is shown in the following screenshot. As long as the shapes are the same, the function will stack this column at the end and calculate the correlation coefficient matrix:

```
In [63]:  np.set_printoptions(suppress= True, linewidth= 125)
          CorrelationCoef_Matrix2 = np.round(np.corrcoef(samples, label, rowvar= False), decimals= 2)
          print("    F1    F2    F3    F4    F5    F6    F7    F8    F9    F10   F11   F12   F13")
          print CorrelationCoef_Matrix2[0:13,13:14].T

               F1    F2    F3    F4    F5    F6    F7    F8    F9    F10   F11   F12   F13
          [[-0.39  0.36 -0.48  0.18 -0.43  0.7  -0.38  0.25 -0.38 -0.47 -0.51  0.33 -0.74]]
```

As you can see, most features have a weak or moderate negative linear relationship, apart from F13. On the other hand, F6 has a strong positive linear relationship. Let's plot this feature and look at this relationship with a scatter plot. The following code block shows three different scatter plots of the features ('RM', 'DIS' and 'LSTAT') and a label column with the help of `matplotlib`:

```
In [64]:  %matplotlib notebook
          %matplotlib notebook
          import matplotlib.pyplot as plt
          from scipy import stats
          fig, (ax1, ax2, ax3) = plt.subplots(1,3 ,figsize= (10,4))
          axs =[ax1,ax2,ax3]
          feature_list = [samples[:,5:6], samples[:,7:8], samples[:,12:13]]
          feature_names = ["RM", "DIS", "LSTAT"]
          for n in range(0, len(feature_list)):
              axs[n].scatter(label, feature_list[n], edgecolors=(0, 0, 0))
              axs[n].set_ylabel(feature_names[n])
              axs[n].set_xlabel('label')
```

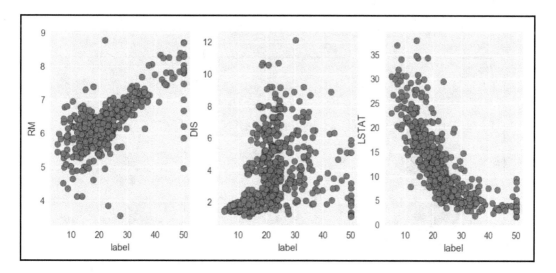

In the preceding code, the RM and label values are dispersed as a positive linear line, which is consistent with the correlation coefficient value of 0.7, as shown in the preceding screenshot. The scatter plot indicates that the higher the RM values of a sample, the higher the label's value. In the middle scatter plot, you can see where the coefficient correlation equals 0.25. This shows a very weak positive linear correlation, which is also visible from the scatter plot. We can conclude that there is no clear relation as the values are dispersed everywhere. The third scatter plot shows a strong linear relation with a correlation coefficient of -0.7. As the LSTAT goes down, the label values start increasing. All of these correlation matrices and scatter plots calculated this with untrimmed data. Let's see how trimming the data by 10% from both sides for each feature and label will change the linear relation results of our dataset:

```
In [65]: %matplotlib notebook
         %matplotlib notebook
         import matplotlib.pyplot as plt
         from scipy import stats
         fig, (ax1, ax2, ax3) = plt.subplots(1,3 ,figsize= (9,4))
         axs = [ax1, ax2, ax3]
         RM_tr = stats.trimboth(samples[:,5:6],0.1)
         label_tr = stats.trimboth(label, 0.1)
         LSTAT_tr = stats.trimboth(samples[:,12:13],0.1)
         DIS_tr = stats.trimboth(samples[:,7:8],0.1)
         feature_names = ["RM", "DIS", "LSTAT"]
         feature_list = [RM_tr, DIS_tr, LSTAT_tr]
         for n in range(0, len(feature_list)):
             axs[n].scatter(label_tr,feature_list[n], edgecolors=(0, 0, 0))
             axs[n].set_ylabel(feature_names[n])
             axs[n].set_xlabel('label')
```

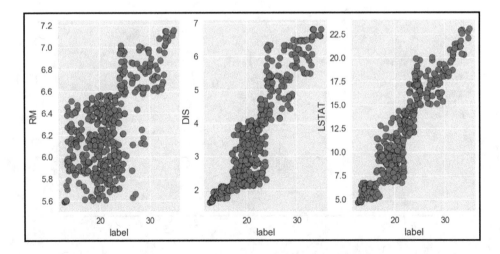

After trimming the data, you can see that all of the features have a strong positive linear correlation with the label, especially the correlation of DIS and LSTAT, where the label has changed tremendously. This shows the power of trimming. You can easily misinterpret your data if you don't know how to deal with outliers. Outliers can change the shape of the distribution and the correlation between other variables, and in the end they can affect your model's performance.

Summary

In this chapter, we covered exploratory data analysis by using the NumPy, SciPy, matplotlib, and Seaborn packages. At the start, we learned how to load and save files and explore our dataset. Then, we explained and calculated important statistical central moments, such as the mean, variance, skewness, and kurtosis. Four important visualizations were used for the graphical representation of univariate and variate analysis, respectively; these were the histogram, box plot, scatter plot, and heatmap. The importance of data trimming was also emphasized using examples.

In the next chapter, we will go one step further and start predicting housing prices using linear regression.

4
Predicting Housing Prices Using Linear Regression

In this chapter, we will introduce supervised learning and predictive modeling by implementing linear regression. In the previous chapter, you learned about exploratory analysis, but haven't looked at modeling yet. In this chapter, we will create a linear regression model to predict housing market prices. Broadly speaking, we are going to predict target variable with the help of its relationship with other variables. Linear regression is very widely used and is a simple model for supervised machine learning algorithms. It's essentially about fitting a line for the observed data. We will start our journey with explaining supervised learning and linear regression. Then, we will analyze the crucial concepts of linear regression such as independent and dependent variables, hyperparameters, loss and error functions, and stochastic gradient descent. For modeling, we will use the same dataset that we used in the previous chapter.

The following topics will be covered in this chapter:

- Supervised learning and linear regression
- Independent and dependent variables
- Hyperparameters
- Loss and error functions
- Implementing our algorithm for a single variable
- Computing stochastic gradient descent
- Using linear regression to model housing prices

Supervised learning and linear regression

Machine learning gives computer systems an ability to learn without explicit programming. One of the most common types of machine learning is supervised learning. Supervised learning consists of a set of different algorithms which formulates a learning problem and solves them by mapping inputs and outputs using historical data. The algorithms analyze the input and a corresponding output, then link them together to find a relationship (learning). Finally, for the new given dataset, it will predict the output by using this learning.

In order to differentiate between supervised and unsupervised learning, we can think about input/output-based modeling. In supervised learning, the computer system will be supervised with labels for every set of input data. In unsupervised learning, the computer system will only use input data without any labels.

As an example, let's assume that we have 1 million photos of cats and dogs. In supervised learning, we label the input data and state whether a given photo is of a cat or a dog. Let's say we have 20 features for each photo (input data). The computer system will know whether the photo is a cat or a dog as it's labeled (output data). When we show the computer system a new photo, it will decide whether it's a cat or a dog by analyzing 20 features of the new photo and make a prediction based on its previous learning. In unsupervised learning, we will just have 1 million cat and dog photos without any labeling stating whether the photo's of a cat or a dog, so the algorithm will cluster the data by analyzing its features without our supervision. After clustering is finished, a new photo will be fed into the unsupervised learning algorithm and the system will tell us which cluster the photo belongs to.

In both scenarios, the system will have a simple or complex decision algorithm. The only difference is whether there is any initial supervision or not. An overview scheme of supervised learning methods is as follows:

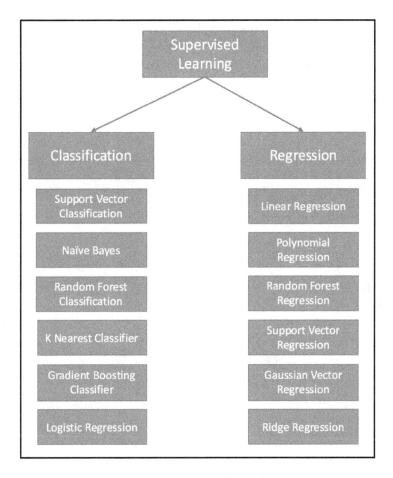

Supervised learning can be split into two types as Classification and Regression, which is shown in the preceding chart. Classification models predict labels. For example, preceding example can be considered a supervised classification problem. In order to perform classification, we need to train classification algorithms such as **Support Vector Classifier** (**SVC**), Random Forests, **k-nearest** neighbors (**KNN**), and so on. Basically, classifiers refer to algorithms that are used to classify (categorize) the data.

Classification methods are used when our target variable is categorical but when our target variable is continuous, regression models are applied, as the aim is predicting numerical value rather than category.

Think about the dataset that we used in the previous chapter: the Boston housing prices dataset. In that dataset, our aim was to analyze feature values statistically because we needed to know how they are distributed, what their basic stats are, and the correlations between each other. In the end, we would like to know how each feature contributes to a housing price. Does it affect positively, negatively, or not at all? If there is a potential effect (relationship) between *feature x* and *housing price A*, how strong or weak is this relation?

We try to answer these questions by building a model for predicting a house price with given features. As a result, after we feed our model with the new features, we expect our model to produce output variables as continuous values (150k, 120, $154, and so on). Let's see a very basic workflow:

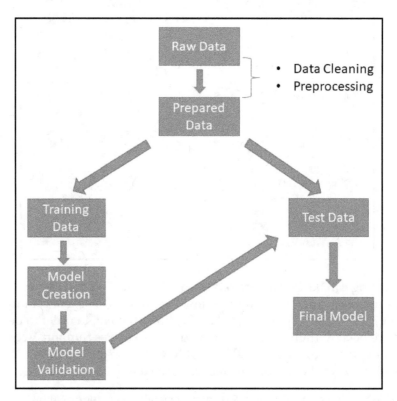

As in the preceding chart, the analysis starts with preparing our data for our model. In this phase, we should clean the data, handle the missing values, and extract the features that will be used. After our data is clean, we should divide it into two parts, training and test data, in order to test the performance of our model.

Important part in model validation is a concept called **overfitting**. In layman's terms, overfitting means learning too much from the training dataset, so our model overfits and produces nearly perfect results for the training dataset. However, it is not flexible enough to produce good results when it comes to data it has never seen, hence it's not able to **generalize** well.

Splitting dataset into training, validation (highly recommended) and test datasets is done to overcome this issue. Training data is where the algorithm initially learns the parameters (weights) of the algorithm and builds a model where errors are minimized. Validation dataset is very useful when you have several algorithms and you need to tune the hyperparameters, or when you have many parameters in your algorithm and need to tune your parameters. The test dataset is used for performance assessment.

In a nutshell, you train your algorithm with training data, then you fine-tune the parameter or weights of your algorithm in the validation dataset and in the last stage, you test your tuned algorithm's performance in the test dataset.

The opposite case to overfitting is underfitting, which means that the algorithm learns less from the data and our algorithm does not fit well with our observations. Let's see graphically what overfitting, underfitting, and best-fitting look like:

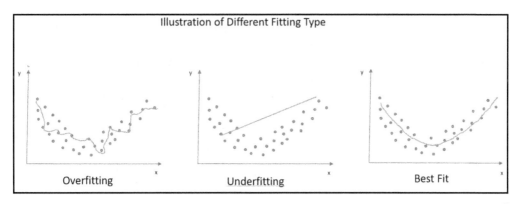

In the preceding graph, as you may have noticed, even though overfitting seems like it fits very well, it generates a regression line which is very unique for this dataset and does not correctly capture the characteristics. The second graph, the underfitting graph, actually couldn't capture the shape of the data; it didn't learn from it and produced a linear regression line when our data is non-linear. The third graph, our best fit graph, fitted very well, grasped the distribution characteristics and produced a curvilinear line. We can expect that it will not have disappointing performance measures for continuation of this dataset.

In this chapter, we will use linear regression as a supervised learning method. We will start by explaining very important concepts such as independent and dependent variables, hyperparameters, and loss and error functions. We will also go through practical examples for linear regression. In the next chapter, we will cover the most important components of the linear regression model: independent and dependent variables.

Independent and dependent variables

As we mentioned in the previous subsection, linear regression is used to predict a value of a variable based on other variables. We are investigating the relationship between input variables, X and the output variable, Y.

In linear regression, **dependent variable** is the variable that we want to predict. The reason that we call it the dependent variable is because of the assumption behind linear regression. The model assumes that these variables depend on the variables that are on the other side of the equation, which are called **independent variables**.

In simple regression model, model will explain how the dependent variable changes based on independent variable.

As an example, let's imagine that we want to analyze how the sales values are effected based on changes in prices for a given product. If you read this sentence carefully, you can easily detect what our dependent and independent variables are. In our example, we assume that the sales value is affected by price changes, in other words, the sales value depends on the price of the product. As a result, the sales value is a dependent value and the price is an independent value. It does not necessarily mean that the price of a given product is not dependent on anything. Of course, it depends on many factors (variables), but in our model, we assume that the price is given, and a given price will change the sales value. The formula for a linear regression line is as follows:

$$Y_i = B_0 + B_1 X_i$$

Where:

Y_i = Estimated value or dependent variable

B_0 = Intercept

B_1 = Slope

X_i = Independent or exploratory variable:

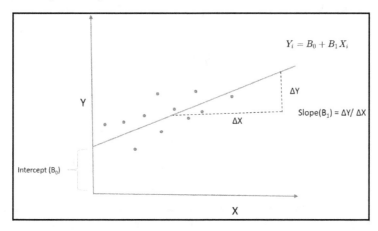

The slope (*B1*) of the linear regression line actually shows the relationship between these two variables. Let's say we calculate the slope as 0.8. This means that a one unit increase in the independent variable is likely to effect a 0.8 unit increase for the estimated value. The preceding linear regression line only generates estimations, which means that they are just predictions of Y for a given X. As you see in the following graph, there is a distance between each observation and the linear line. That distance is called an **error**, which is expected and also very important in regression line fitting and model evaluation:

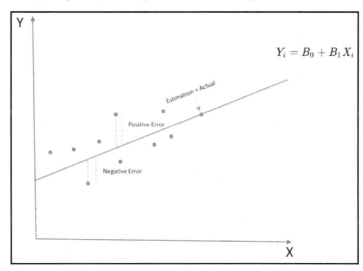

The most common way of fitting a linear regression line is by using the **least squares** method. This method fits the regression line by minimizing the sum of the squares of these errors. The formula is as follows:

$$\sum (y - \hat{y})^2$$

The reason that these errors are squared is that we don't want negative and positive errors to cancel each other out. In the model evaluation, R-squared, F-test, and **Root Mean Square Error (RMSE)** are used. They all use **Sum of Squares Total (SST)** and **Sum of Squares Error (SSE)** as base measures. As you can see, in SSE, we once again calculate the difference between predicted values and actual values, take the square, and sum them up in order to evaluate how well the regression line fits the data.

As we mentioned previously, the least squares method aims to minimize squared errors (residuals) and find the slope and intercept value which will fit best to the data points. As this is a closed-form solution, you can easily calculate it by hand in order to see what this method does in the background. Let's use an example with a small dataset:

Milk Consumption (liter per week)	Height
14	175
20	182
10	170
15	185
12	164
15	173
22	181
25	193
12	160
13	165

Let's assume that we have 10 observations for weekly milk consumption and the height values for the people who consume the milk, as per the preceding table. If we plot this data, we can see that there is a positive correlation between daily milk consumption and height:

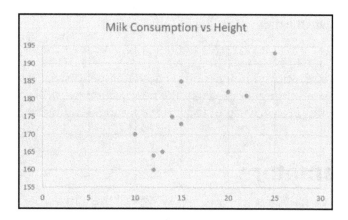

Now, we want to fit the linear regression line by using the least squares method, which is done by using the following formula for the slope and intercept:

$$B_0 = \frac{(\sum y)(\sum x^2) - (\sum x)(\sum xy)}{n(\sum x^2) - (\sum x)^2}$$

$$B_1 = \frac{n(\sum xy) - (\sum x)(\sum y)}{n \sum x^2 - (\sum x)^2}$$

Then, we need to create a table to help us work out calculations such as for the sum of x, y, xy, x^2, and y^2:

	x (Milk Consumption)	y (Height)	xy	x^2	y^2
1	14	175	2,450	196	30,625
2	20	182	3,640	400	33,124
3	10	170	1,700	100	28,900
4	15	185	2,775	225	34,225
5	12	164	1,968	144	26,896
6	15	173	2,595	225	29,929
7	22	181	3,982	484	32,761
8	25	193	4,800	625	37,249
9	12	160	1,920	144	25,600
10	13	165	2,145	169	27,225
$\sum$	158	1,748	27,975	2,712	306,534

B_0= (1748*2712)-(158*27975) / (10*2712)-(158)^2 = 320526/2156 = 148.66

B_1= (10*27975)-(158*1748) / (10*2712)-(158)^2 =1.65

Then, we can formulate the regression line as follows:

$$Y = 148.66 + 1.65X$$

In this subsection, we mentioned independent and dependent variables and introduced the linear regression line and fitting methods. In the next section, we will cover hyperparameters, which are very useful in regression model tuning.

Hyperparameters

Before we start, maybe it's better to explain why we call them hyperparameters and not parameters. In machine learning, model parameters can be learned from the data, which means that while you train your model, you fit the model's parameters. On the other hand, we usually set hyperparameters before we start training the model. In order to give an example, you can think of coefficients in regression models as model parameters. A hyperparameter example, we can say the learning rate in many different models or the number of clusters (k) in k-means clustering.

Another important thing is the relationship between model parameters and hyperparameters, and how they shape our machine learning model, in other words, the hypothesis of our model. In machine learning, parameters are used for configuring the model, and this configuration will tailor the algorithm specifically for our dataset. What we need to handle is how to optimize the hyperparameters. In addition, this optimization may be performed during validation as previously mentioned. Optimizing hyperparameters will bring performance improvement in many cases.

You can also think of hyperparameters as high-level parameters on top of model parameters. Imagine a case where you use k-means clustering which is unsupervised learning. If you use the wrong cluster number (K) as a hyperparameter, it's guaranteed that you cannot have the suitable fit for your data.

By now, you should be asking how we can tune the hyperparameters if we set them manually prior to training the model. There are several ways to tune hyperparameters. The bottom line for this optimization is to test the algorithm with a set of different hyperparameters and calculate the error function or loss function for each scenario before picking the hyperparameter set which has better performance.

In this section, we briefly covered parameters, hyperparameters, and their differences. In the next section, we will touch on loss and error functions, which are very important for hyperparameter optimization.

Loss and error functions

In the previous subsections, we explain supervised and unsupervised learning. Regardless of which machine learning algorithm is used, our main challenge is regarding issues with optimization. In optimization functions, we are actually trying to minimize the loss function. Imagine a case where you are trying to optimize your monthly savings. In a closed state, what you will do is minimize your spending, in other words, minimize your loss function.

A very common way to build a loss function is starting with the difference between the predicted value and the actual value. In general, we try to estimate the parameters of our model, and then prediction is made. The main measurement that we can use to evaluate how good our prediction is involves calculating the difference between the actual values:

$$L(x) = p - \hat{p}$$

In different models, different loss functions are used. For example, you can use a mean squared error in your regression model but it's probably not a good idea to use it as a loss function for your classification model. As an example, you can calculate the mean squared error as follows:

$$MSE = \frac{1}{N} \sum_{i=1}^{n} (y_i - (\beta_1 x_i + \beta_0))^2$$

Where the regression model is as follows:

$$Y_i = \beta_0 + \beta_1 X_i$$

There are many different loss functions that can be used in different machine learning models. Here are some important ones with a brief explanation and their usage:

Loss Function	Explanation
Cross-Entropy	This is used for classification models where the output of the model is a probability between 0-1. It's a logarithmic loss function, also known as log loss. When the predicted probability approaches 1.0, which is a perfect model, cross-entropy loss decreases.
MAE(L1)	Calculates the average errors. As it just uses the absolute value, it does not amplify weights to large errors. This is useful when the large errors are tolerable compared to small ones.
MSE(L2)	Takes the square root of errors. Amplifies the weights of large errors. This is useful when the large errors are undesirable.

Hinge	This is a loss function used for linear classifier models such as Support Vector Machines.
Huber	This is a loss function for regression models. It is very similar to MSE but less sensitive to outliers.
Kullback-Leibler	Kullback-Leibler divergence measures the difference between two probability distributions. The KL loss function used a lot in t-distributed stochastic neighbor embedding algorithms.

In machine learning algorithms, loss functions are crucial while updating the weights of variables. Let's say you use backpropagation to train neural networks. In each iteration, the total error is calculated. Then, the weights are updated in order to minimize the total error. Therefore, using the correct loss function directly affects the performance of your machine learning algorithm as it has a direct effect on model parameters. In the next chapter, we will start simple linear regression with a single variable in housing data.

Univariate linear regression with gradient descent

In this subsection, we will implement univariate linear regression for the Boston housing dataset, which we used for exploratory data analysis in the previous chapter. Before we fit the regression line, let's import the necessary libraries and load the dataset as follows:

```
In [1]: import numpy as np
        import pandas as pd
        from sklearn.cross_validation import train_test_split
        from sklearn.linear_model import LinearRegression
        import matplotlib.pyplot as plt
        %matplotlib inline
In [2]: from sklearn.datasets import load_boston
        dataset = load_boston()
        samples , label, feature_names = dataset.data, dataset.target,
dataset.feature_names
In [3]: bostondf = pd.DataFrame(dataset.data)
        bostondf.columns = dataset.feature_names
        bostondf['Target Price'] = dataset.target
        bostondf.head()
Out[3]:    CRIM    ZN    INDUS CHAS  NOX   RM    AGE   DIS    RAD   TAX
PTRATIO  B      LSTAT  Target Price
        0 0.00632 18.0  2.31  0.0  0.538 6.575 65.2  4.0900 1.0   296.0
15.3  396.90  4.98       24.0
        1 0.02731 0.0   7.07  0.0  0.469 6.421 78.9  4.9671 2.0   242.0
```

```
17.8  396.90   9.14      21.6
         2 0.02729  0.0   7.07  0.0  0.469  7.185 61.1  4.9671 2.0  242.0
17.8  392.83   4.03      34.7
         3 0.03237  0.0   2.18  0.0  0.458  6.998 45.8  6.0622 3.0  222.0
18.7  394.63   2.94      33.4
         4 0.06905  0.0   2.18  0.0  0.458  7.147 54.2  6.0622 3.0  222.0
18.7  396.90   5.33      36.2
```

Compared to the previous chapter, we are using the Pandas DataFrame instead of the numpy array in order to show you the usage of dataframe as it`s a very convenient data structure as well. Technically, it doesn't make any difference whether you store the data in a numpy array or a Pandas DataFrame in most cases if you have only numerical values. Let's add the target value to our data frame and see the relationship between the RM feature and the target value with a scatter plot:

```
In [4]: import matplotlib.pyplot as plt
        bostondf.plot(x='RM', y='Target Price', style= 'o')
        plt.title('RM vs Target Price')
        plt.ylabel('Target Price')
        plt.show()
```

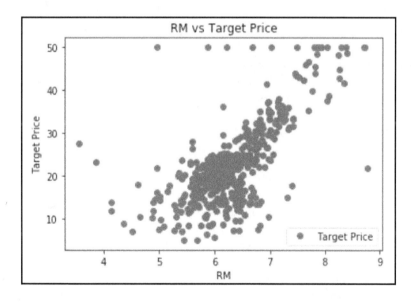

As you can see from the plot, there is a positive correlation between the average number of rooms per dwelling (RM) and the house price, as expected. Now, we will see the magnitude of this relation and try to predict the housing prices by using this relation.

Imagine that you have very limited knowledge of what linear regression is. Let's say that you just have familiarity with the equation but you don't know what an error function is, why we need iteration, what gradient descent is, and why we use it in linear regression models. In this case, what you would do is simply start passing some initial values for the coefficient and intercept to the equation before calculating the predicted value.

After you calculate several predicted values, you would compare them with the actual values and see how far you are from the reality. The next step would be to change the coefficient or intercept, or do both to see whether you can get closer results to the actual values. If you feel comfortable with that process, that's what our algorithm will do in smarter way.

In order to better understand the linear regression model steps, we separate the code into several blocks. First, let's create a function which returns the prediction value as a result of the regression line:

```
In [5]: def prediction(X, coefficient, intercept):
            return X*coefficient + intercept
```

The preceding function computes a linear regression model as follows:

$$Y_i = \beta_1 X_i + \beta_0$$

Then, we need a cost function, which will be calculated in each iteration. As a `cost_function`, we will use the mean squared error, which will be the average of the total squared difference between predictions and actual values:

```
In [6]: def cost_function(X, Y, coefficient, intercept):
            MSE = 0.0
            for i in range(len(X)):
                MSE += (Y[i] - (coefficient*X[i] + intercept))**2
            return MSE / len(X)
```

Our last code block will be for updating the weights. When we talk about the weights, it's not only about coefficients of independent variables but also the intercept. Intercept is also known as bias. In order to update the weights logically, we need an iterative optimization algorithm which finds the minimum value of a given function. In this example, we will use the gradient descent method to minimize the loss function in each iteration. Let's discover what gradient descent does step by step.

First of all, we should initialize the weights (intercept and coefficient) and calculate the mean squared error. Then, we need to calculate the gradient, which means looking at how the mean squared error changes when we change the weights.

In order to change the weights in a smarter way, we need to understand which direction we have to change our coefficient and intercept. This means we should calculate the gradient of the error function when we change the weights. We can calculate the gradient by taking the partial derivative of a loss function for coefficients and intercept.

In univariate linear regression, we have only one coefficient. After we calculate the partial derivatives, the algorithm will adjust the weights and recalculate the mean squared error. This process will iterate until the updated weight does not reduce the means squared error anymore:

```
In [7]: def update_weights(X, Y, coefficient, intercept, learning_rate):
            coefficient_derivative = 0
            intercept_derivative = 0
            for i in range(len(X)):
                coefficient_derivative += -2*X[i] * (Y[i] -
(coefficient*X[i] + intercept))
                intercept_derivative += -2*(Y[i] - (coefficient*X[i] +
intercept))
            coefficient -= (coefficient_derivative / len(X)) *
learning_rate
            intercept -= (intercept_derivative / len(X)) * learning_rate
            return coefficient, intercept
```

The preceding code block defines the function for updating the weight and then returns the updated coefficient and intercept. Another important parameter in this function is `learning_rate`. This learning rate will decide the magnitude of the change:

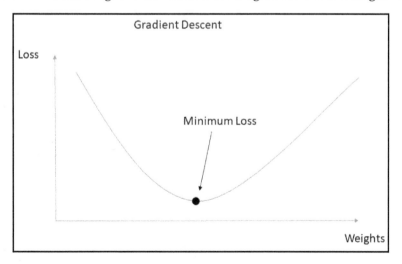

In the preceding graph, you can see that our loss function has the shape of $x = y^2$ as it's the sum of the squared error. With gradient descent, we are trying to find the minimum loss, as shown in the graph, where the partial derivatives are very close to zero. As we mentioned previously, we start our algorithm by initializing the weights, which means that we start from the points which are far from the minimum. In each iteration, we will update the weight, which decreases the loss. This means that given enough iterations, we will converge to the global minimum. The learning rate will decide how fast this convergence will happen.

In other words, a high learning rate will look like a huge jump from one point to another when we update the weights. A low learning rate will approximate to the global minimum (the desired minimum loss point) slowly. As the learning rate is a hyperparameter, we need to set it before we run it:

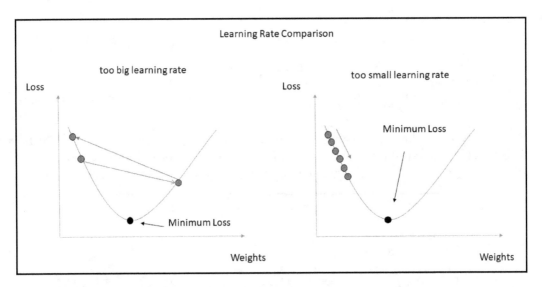

So, do we need to set a big or small learning rate? The answer is that we should find the optimal rate. If we set a big learning rate, our algorithm will be overshoot the minimum. We can easily miss the minimum as these jumps will never let our algorithm converge to the global minimum. On the other hand, if we set the learning rate too small, we may need a lot of iterations in order to converge.

Having previous code blocks, it's time to write the main function. The main function should follow the flow here like we discussed previously:

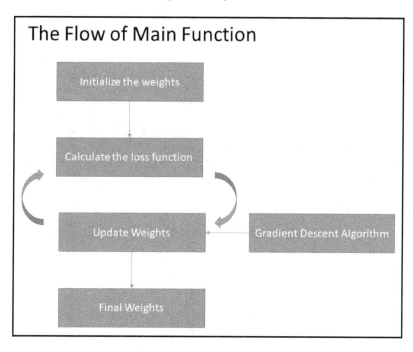

Then, the code block for the main function should be as follows:

```
In [8]: def train(X, Y, coefficient, intercept, LearningRate, iteration):
            cost_hist = []
            for i in range(iteration):
                coefficient, intercept = update_weights(X, Y, coefficient,
intercept, learning_cost =                     cost_function(X, Y,
coefficient, intercept)
                cost_hist.append(cost)
            return coefficient, intercept, cost_hist
```

Now, we have all of the code blocks defined and we are ready to run our univariate model. Before we run the `train()` function, we need to set the hyperparameters and initial values for the coefficient and intercept. You can create the variables, set the values and then put these variables as parameters of the function, as follows:

```
In [9]: learning_rate = 0.01
        iteration = 10001
        coefficient = 0.3
        intercept = 2
```

```
X = bostondf.iloc[:, 5:6].values
Y = bostondf.iloc[:, 13:14].values

coefficient, intercept, cost_history = train(X, Y, coefficient,
intercept, learning_rate, iteration)
```

Or, you can pass these values as keyword arguments when you call the function:

```
coefficient, intercept, cost_history = train(X, Y, coefficient, intercept =
2, learning_rate = 0.01, iteration = 10001)
```

Our main function will return two arrays, which gives us the final values of intercept and coefficient. Additionally, the main function will return a list of loss values, which is the result of each iteration. These are the results of the mean squared error in each iteration. It is very useful to have this list in order to track how loss is changing in each iteration:

```
In [10]: coefficient
Out[10]: array([8.57526661])
In [11]: intercept
Out[11]: array([-31.31931428])
In [12]: cost_history
         array([54.18545801]),
       array([54.18036786]),
       array([54.17528017]),
       array([54.17019493]),
       array([54.16511212]),
       array([54.16003177]),
       array([54.15495385]),
       array([54.14987838]),
       array([54.14480535]),
       array([54.13973476]),
       array([54.13466661]),
       array([54.12960089]),
       array([54.12453761]),
       array([54.11947677]),
       array([54.11441836]),
       array([54.10936238]),
       array([54.10430883]),
       ...]
```

As you see from the coefficient value, that one unit increase in RM increases the housing price to around $8,575. Now, let's calculate the predicted values by injecting the calculated intercept and coefficient into the regression formula. Then, we can plot the linear regression line and see how it fits our data:

```
In [13]: y_hat = X*coefficient + intercept
         plt.plot(X, Y, 'bo')
         plt.plot(X, y_hat)
         plt.show()
```

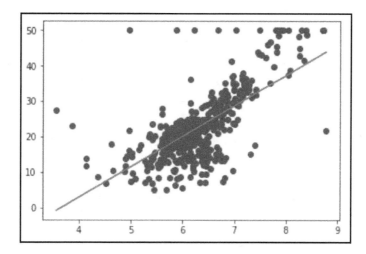

In this section, we applied the univariate model by picking a single variable. In the following subsections, we will perform a multivariate linear regression model by adding more independent variables into our model, which means that we will have several coefficients to optimize for the best fit.

Using linear regression to model housing prices

In the section, we will perform multivariate linear regression for the same dataset. In contrast to the previous section, we will use the sklearn library to show you several ways of performing linear regression models. Before we start the linear regression model, we will trim the dataset proportionally from both sides by using the `trimboth()` method. By doing this, we will cut off the outliers:

```
In [14]: import numpy as np
```

```
            import pandas as pd
            from scipy import stats
            from sklearn.cross_validation import train_test_split
            from sklearn.linear_model import LinearRegression
In [15]: from sklearn.datasets import load_boston
            dataset = load_boston()
In [16]: samples , label, feature_names = dataset.data, dataset.target,
dataset.feature_names
In [17]: samples_trim = stats.trimboth(samples, 0.1)
            label_trim = stats.trimboth(label, 0.1)
In [18]: print(samples.shape)
            print(label.shape)
            (506, 13)
            (506,)
In [19]: print(samples_trim.shape)
            print(label_trim.shape)
            (406, 13)
            (406,)
```

As you see in the preceding code block, our sample size has decreased from 506 to 406 for each attribute and label column as we trimmed the data on the left and right by 10%:

```
In [20]: from sklearn.model_selection import train_test_split
            samples_train, samples_test, label_train, label_test =
train_test_split(samples_trim, In [21]:        In [21]:
print(samples_train.shape)
            print(samples_test.shape)
            print(label_train.shape)
            print(label_test.shape)
            (324, 13)
            (82, 13)
            (324,)
            (82,)
In [22]: regressor = LinearRegression()
            regressor.fit(samples_train, label_train)
Out[22]: LinearRegression(copy_X=True, fit_intercept=True, n_jobs=1,
normalize=False)
In [23]: regressor.coef_
Out[23]: array([ 2.12924665e-01, 9.16706914e-02, 1.04316071e-01,
-3.18634008e-14,
                  5.34177385e+00, -7.81823481e-02, 1.91366342e-02,
2.81852916e-01,
                  3.19533878e-04, -4.24007416e-03, 1.94206366e-01,
3.96802252e-02,
                  3.81858253e-01])
In [24]: regressor.intercept_
Out[24]: -6.899291747292615
```

We then use the `train_test_split()` method to split our dataset into train and test. This approach is very common in machine learning algorithms. You divide your data into two sets and let your model train (learn) before testing your model with the other part of the data. The reason for using this approach is to decrease the bias while you are validating your model. Each coefficient represents how much the target value is expected to change when we increase these samples by a single unit:

```
In [25]: label_pred = regressor.predict(samples_test)
In [26]: plt.scatter(label_test, label_pred)
         plt.xlabel("Prices")
         plt.ylabel("Predicted Prices")
         plt.title("Prices vs Predicted Prices")
         plt.axis("equal")
Out[26]: (11.770143369175626, 34.22985663082437, 10.865962968036989,
34.20549738482051)
```

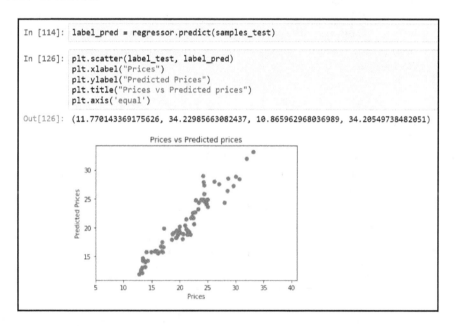

Let's test our model in the preceding code block. In order to run our model on a dataset, we can use the `predict()` method. This method will run the given dataset on our model and return the results. Ideally, we are expecting the point to be distributed as an $x=y$ line shape, in other words, such as a linear line which has a 45 degree angle. This is a perfect, 1-on-1 matching prediction. We cannot say that our prediction is perfect, but just by looking at the scatter plot, we can conclude that the prediction is not a bad one. We can look at two important metrics for the model's performance as follows:

```
In [27]: from sklearn.metrics import mean_squared_error
         from sklearn.metrics import r2_score
         mse = mean_squared_error(label_test, label_pred)
         r2 = r2_score(label_test, label_pred)
         print(mse)
         print(r2)
         2.032691267250478
         0.9154474686142619
```

The first one is the mean squared error. You may have noticed that it decreases significantly when we add more independent variables into the model. The second one is R^2, which is coefficient of determination, or in other words, the regression score. The coefficient of determination explains how much variance in a dependent variable is explained by your independent variables. In our model, the result is 0.91, which explains that 91% of the variance in housing prices can be explained by our thirteen features. As you can see, sklearn has many useful built-in functions for linear regression which may speed up your model building process. On the other hand, you can also easily build your linear regression model just by using numpy, which might give you more flexibility as you can control each part of your algorithm.

Summary

Linear regression is one of the most common techniques for modeling the relationship between continuous variables. The application of this method is very widely used in the industry. We started modeling part of the book on linear regression, not just because it's very popular, but because it's a relatively easy technique and contains most of the elements which almost every machine learning algorithm has.

In this chapter, we learned about supervised and unsupervised learning and built a linear regression model by using the Boston housing dataset. We touched upon different important concepts such as hyperparameters, loss functions, and gradient descent. The main purpose of this chapter was to give you sufficient knowledge so that you can build and tune a linear regression model and understand what it does step by step. We looked at two practical cases where we used univariate and multivariate linear regression. You have also experienced the usage of numpy and sklearn in linear regression. We highly encourage that you practice this further with different datasets and examine how the outcome changes when you change your hyperparameters.

In the next chapter, we will learn about the clustering method and practice it with an example on wholesale distributor dataset.

5
Clustering Clients of a Wholesale Distributor Using NumPy

You are definitely advancing your skills by seeing NumPy in action for various use cases. This chapter is about a different type of analysis than what you have seen so far. Clustering is an unsupervised learning technique that is used for understanding and capturing the various formations in your dataset. Since you don't have label to supervise your learning algorithm, in many cases, visualization is the key, which is why you will see various visualization techniques as well.

In this chapter, we will cover the following topics:

- Unsupervised learning and clustering
- Hyperparameters
- Extending simple algorithm to cluster the clients of a wholesale distributor

Unsupervised learning and clustering

Let's quickly review supervised learning with an example. When you are training machine-learning algorithms, you are able to observe and direct the learning by providing labels. Think about the following dataset, where each row indicates a customer and each column represents a different feature such as **Age**, **Gender**, **Income**, **Profession**, **Tenure** and **City**. Take a look at this table:

	Age	Gender	Income	Profession	Tenure	City
Records	35	M	60,000	IT	12	KRK
	23	F	90,000	Sales	3	WAW
	18	M	12,000	Student	1	KRK
	42	F	128,000	Doctor	13	KRK
	34	M	63,000	Manager	8	WAW
	56	M	82,000	Teacher	30	WAW

You may want to perform different kinds of analysis. One of them could be to predict which of the customers is likely to leave, namely, churn analysis. To do that, you need to label each customer based on their history to indicate which customers have left or stayed, as displayed here, in this table:

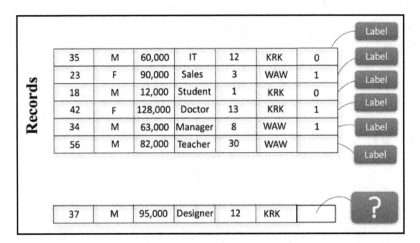

Your algorithm will learn the characteristics of customers based on their label. Algorithm will learn the characteristics of customers who left or stayed, and, when you would like to score a new customer based on these features, algorithm will make its predictions based on that learning. This is called **supervised learning**.

Another question which could be asked about this dataset could be this: How many customer groups that are present in this dataset so that every customer within its group is similar to others and dissimilar to customers which belong to other groups. Popular clustering algorithm such as k-means can help you figure that out. For example once k-means assign customers to different clusters, one cluster can mostly include customers who are less than 30 years old and have **IT** as profession and another cluster can mostly include customers who are above 60 years old and has **Teacher** as profession. You don't need to label your dataset to perform this analysis, as it's enough for the algorithm to see the records and work out the similarity between them. This type of learning is called **unsupervised learning** since there is no supervision.

It's helpful to visualize the dataset first when you conduct such analysis. You can start with available datasets to build your processing and modeling workflow. The following code snippets show you how to visualize three-dimensional datasets with `plotly`. `plotly` is a library that allows you to draw many different interactive charts for exploratory analysis, and makes data exploration easier.

First, you need to install the necessary libraries with the following snippet:

```
# Installing necessary libraries with pip
!pip install plotly --user
!pip install cufflinks -user
```

Then, you import the necessary libraries, using this code:

```
# Necessary imports
import os
import sys
import numpy as np
import pandas
import matplotlib.pyplot as plt
%matplotlib inline
import plotly.plotly as py
from plotly.offline import download_plotlyjs, init_notebook_mode, plot,
iplot
import cufflinks as cf
import plotly.graph_objs as go

init_notebook_mode(connected=True)
sys.path.append("".join([os.environ["HOME"]]))
```

You will use the `iris` dataset, which is available in the `sklearn.datasets` module, as shown here:

```
from sklearn.datasets import load_iris
iris_data = load_iris()
```

`iris` data has four features; they are as follows:

```
iris_data.feature_names

['sepal length (cm)',
 'sepal width (cm)',
 'petal length (cm)',
 'petal width (cm)']
```

First, let's check the first two features, which are these:

```
x = [v[0] for v in iris_data.data]
y = [v[1] for v in iris_data.data]
```

Create a `trace`, then data and the figure, as shown here:

```
trace = go.Scatter(
    x = x,
    y = y,
    mode = 'markers'
)

layout= go.Layout(
    title= 'Iris Dataset',
    hovermode= 'closest',
    xaxis= dict(
        title= 'sepal length (cm)',
        ticklen= 5,
        zeroline= False,
        gridwidth= 2,
    ),
    yaxis=dict(
        title= 'sepal width (cm)',
        ticklen= 5,
        gridwidth= 2,
    ),
    showlegend= False
)
```

```
data = [trace]

fig= go.Figure(data=data, layout=layout)
plot(fig)
```

This will give you the following output, as shown in this diagram:

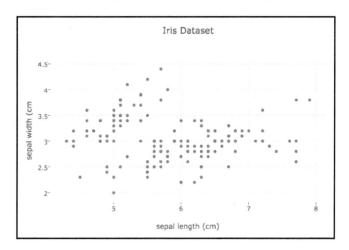

You can keep looking at other variables, but to better understand relationships between features in one chart, you can use the scatterplot matrix. Creating a pandas.DataFrame to use with plotly is more convenient here:

```
import pandas as pd
df = pd.DataFrame(iris_data.data,
columns=['sepal length (cm)',
'sepal width (cm)',
'petal length (cm)',
'petal width (cm)'])

df['class'] = [iris_data.target_names[i] for i in iris_data.target]

df.head()
```

	sepal length (cm)	sepal width (cm)	petal length (cm)	petal width (cm)	class
0	5.1	3.5	1.4	0.2	setosa
1	4.9	3.0	1.4	0.2	setosa
2	4.7	3.2	1.3	0.2	setosa
3	4.6	3.1	1.5	0.2	setosa
4	5.0	3.6	1.4	0.2	setosa

Using the `plotly` figure factory, you can plot `scatterplot` matrix, as shown here:

```
import plotly.figure_factory as ff

fig = ff.create_scatterplotmatrix(df, index='class', diag='histogram',
size=10, height=800, width=800)

plot(fig)
```

This will give you the following plot, as shown in the following diagram:

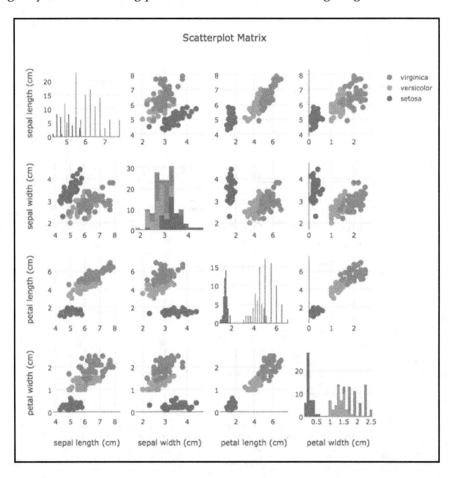

At first look, **petal length**, **petal width**, and **sepal length** seem to be good candidates for modeling. You can use 3D charts to inspect this dataset further, by using this code:

```
# Creating data for the plotly
trace1 = go.Scatter3d(
    # Extracting data based on label
    x=[x[0][0] for x in zip(iris_data.data, iris_data.target) if x[1] ==
0],
    y=[x[0][2] for x in zip(iris_data.data, iris_data.target) if x[1] ==
0],
    z=[x[0][3] for x in zip(iris_data.data, iris_data.target) if x[1] ==
0],
    mode='markers',
    marker=dict(
        size=12,
        line=dict(
            color='rgba(217, 217, 217, 0.14)',
            width=0.5
        ),
        opacity=0.8
    )
)

trace2 = go.Scatter3d(
    # Extracting data based on label
    x=[x[0][0] for x in zip(iris_data.data, iris_data.target) if x[1] ==
1],
    y=[x[0][2] for x in zip(iris_data.data, iris_data.target) if x[1] ==
1],
    z=[x[0][3] for x in zip(iris_data.data, iris_data.target) if x[1] ==
1],
    mode='markers',
    marker=dict(
        color='rgb(#3742fa)',
        size=12,
        symbol='circle',
        line=dict(
            color='rgb(204, 204, 204)',
            width=1
        ),
        opacity=0.9
    )
)

trace3 = go.Scatter3d(
    # Extracting data based on label
    x=[x[0][0] for x in zip(iris_data.data, iris_data.target) if x[1] ==
```

```
2],
    y=[x[0][2] for x in zip(iris_data.data, iris_data.target) if x[1] ==
2],
    z=[x[0][3] for x in zip(iris_data.data, iris_data.target) if x[1] ==
2],
    mode='markers',
    marker=dict(
        color='rgb(#ff4757)',
        size=12,
        symbol='circle',
        line=dict(
            color='rgb(104, 74, 114)',
            width=1
        ),
        opacity=0.9
    )
)


data = [trace1, trace2, trace3]

# Layout settings
layout = go.Layout(
    scene = dict(
        xaxis = dict(
            title= 'sepal length (cm)'),
        yaxis = dict(
            title= 'petal length (cm)'),
        zaxis = dict(
            title= 'petal width (cm)'),),
)


fig = go.Figure(data=data, layout=layout)
plot(fig)
```

This will give you the following plot, which is interactive, as shown in this diagram:

Interactive plotting of petal length, petal width, and sepal length

Using these charts, you can understand your data better and prepare for modeling.

Hyperparameters

Hyperparameter could be considered as high-level parameter which determines one of the various properties of a model such as complexity, training behavior and learning rate. These parameters naturally differ from model parameters as they need to be set before training starts.

For example, the k in k-means or k-nearest-neighbors is a hyperparameter for these algorithms. The k in k-means denotes the number of clusters to be found, and the k in k-nearest-neighbors denotes the number of closest records to be used to make predictions.

Tuning hyperparameters is a crucial step in any machine learning project to improve predictive performance. There are different techniques for tuning, such as grid search, randomized search and bayesian optimization, but these techniques are beyond the scope of this chapter.

Let's have a quick look at the k-means algorithms parameters from the scikit-learn library via this screenshot:

```
In [1]: from sklearn.cluster import KMeans
        KMeans?

        Init signature: KMeans(n_clusters=5, init='k-means++', n_init=15, max_iter=300, tol=0.0001, precompute_distances='auto', verbose=0, random_state=None, copy_x=True, n_jobs=1, a
        lgorithm='auto')
        Docstring:
        K-Means clustering

        Read more in the :ref:'User Guide <k_means>'.

        Parameters
        ----------

        n_clusters : int, optional, default: 8
            The number of clusters to form as well as the number of
            centroids to generate.

        init : {'k-means++', 'random' or an ndarray}
            Method for initialization, defaults to 'k-means++':

            'k-means++' : selects initial cluster centers for k-mean
            clustering in a smart way to speed up convergence. See section
            Notes in k_init for more details.

            'random': choose k observations (rows) at random from data for
            the initial centroids.

            If an ndarray is passed, it should be of shape (n_clusters, n_features)
            and gives the initial centers.

        n_init : int, default: 10
            Number of time the k-means algorithm will be run with different
            centroid seeds. The final results will be the best output of
            n_init consecutive runs in terms of inertia.

        max_iter : int, default: 300
            Maximum number of iterations of the k-means algorithm for a
            single run.

        tol : float, default: 1e-4
            Relative tolerance with regards to inertia to declare convergence
```

As you can see, there are many parameters to play with, and you should at least look at the function signature of algorithms to see the options you have before running algorithms.

Let's play with some of them. The baseline model will work with the sample data, with almost default settings, as shown here:

```
from sklearn.datasets.samples_generator import make_blobs
X, y = make_blobs(n_samples=20, centers=3, n_features=3, random_state=42)

k_means = KMeans(n_clusters=3)
y_hat = k_means.fit_predict(X)
```

y_hat is keeping the membership information of clusters, and this is the same with the original labels, as you can see here:

```
y_hat
# array([0, 2, 1, 1, 1, 0, 2, 0, 0, 0, 2, 0, 1, 2, 1, 2, 2, 1, 0, 1],
dtype=int32)

y
# array([0, 2, 1, 1, 1, 0, 2, 0, 0, 0, 2, 0, 1, 2, 1, 2, 2, 1, 0, 1])
```

You can play with different options to see how it will affect the training and predictions.

The loss function

The loss function helps algorithms to update model parameters during training through measuring the error, which is an indication of predictive performance. Loss function is usually denoted as follows:

$$L(w) = p - \hat{p}$$

Where L measures the difference between the prediction and the actual value. During the training process, this error is minimized. Different algorithms have different loss functions, and the number of iterations will depend on convergence conditions.

For example, the loss function for k-means minimizes the square distances between a points and closest cluster mean as follows:

$$L = \sum_{k=1}^{K} \sum_{i=1}^{n} ||x_i - \mu_k||^2$$

You will see detailed implementation in the following section.

Implementing our algorithm for a single variable

Let's implement the k-means algorithm for a single variable. You will start with one dimensional vector, which has 20 records, as shown here:

```
data = [1,2,3,2,1,3,9,8,11,12,10,11,14,25,26,24,30,22,24,27]

trace1 = go.Scatter(
    x=data,
    y=[0 for x in data],
    mode='markers',
    name='Data',
    marker=dict(
        size=12
    )
)

layout = go.Layout(
```

```
title='1D vector',
)

traces = [trace1]

fig = go.Figure(data=traces, layout=layout)

plot(fig)
```

This will output following plot, as shown in this diagram:

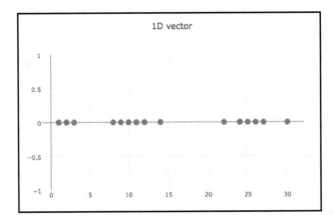

Our aim is to find 3 clusters which are visible in the data. In order to start implementing the k-means algorithm, you need to initialize cluster centers by choosing random indexes, as shown here:

```
n_clusters = 3

c_centers = np.random.choice(X, n_clusters)

print(c_centers)

# [ 1 22 26]
```

Next, you need to calculate distances between each point and the cluster centers, so use this code:

```
deltas = np.array([np.abs(point - c_centers) for point in X])

deltas
array([[ 7, 26, 10],
       [ 6, 25,  9],
       [ 5, 24,  8],
```

```
       [ 6, 25,   9],
       [ 7, 26,  10],
       [ 5, 24,   8],
       [ 1, 18,   2],
       [ 0, 19,   3],
       [ 3, 16,   0],
       [ 4, 15,   1],
       [ 2, 17,   1],
       [ 3, 16,   0],
       [ 6, 13,   3],
       [17,  2,  14],
       [18,  1,  15],
       [16,  3,  13],
       [22,  3,  19],
       [14,  5,  11],
       [16,  3,  13],
       [19,  0,  16]])
```

Now cluster membership can be calculated, by using this code:

```
deltas.argmin(1)
# array([0, 0, 0, 0, 0, 0, 0, 0, 2, 2, 2, 2, 2, 1, 1, 1, 1, 1, 1, 1])
```

Now you need to calculate distances between records and cluster centers, by using this code:

```
c_centers = np.array([X[np.where(deltas.argmin(1) == i)[0]].mean() for i in
range(3)])

print(c_centers)

# [ 3.625      25.42857143 11.6       ]
```

This was one iteration; you could continue calculating new cluster centers until there is no improvement.

You can write a function to wrap all this functionality, as shown here:

```
def Kmeans_1D(X, n_clusters, random_seed=442):

    # Randomly choose random indexes as cluster centers
    rng = np.random.RandomState(random_seed)
    i = rng.permutation(X.shape[0])[:n_clusters]
    c_centers = X[i]

    # Calculate distances between each point and cluster centers
    deltas = np.array([np.abs(point - c_centers) for point in X])
```

```
    # Get labels for each point
    labels = deltas.argmin(1)
    while True:
      # Calculate mean of each cluster
      new_c_centers = np.array([X[np.where(deltas.argmin(1) == i)[0]].mean()
for i in range(n_clusters)])

      # Calculate distances again
      deltas = np.array([np.abs(point - new_c_centers) for point in X])

      # Get new labels for each point
      labels = deltas.argmin(1)

      # If there's no change in centers, exit
      if np.all(c_centers == new_c_centers):
        break
      c_centers = new_c_centers

   return c_centers, labels

c_centers, labels = Kmeans_1D(X, 3)

print(c_centers, labels)

# [11.16666667 25.42857143  2.85714286] [2 2 2 2 2 2 0 0 0 0 0 0 0 1 1 1 1
1 1 1]
```

Let's chart the cluster centers, using this code:

```
trace1 = go.Scatter(
    x=X,
    y=[0 for num in X],
    mode='markers',
    name='Data',
    marker=dict(
    size=12
    )
)

trace2 = go.Scatter(
    x = c_centers,
    y = [0 for num in X],
    mode='markers',
    name = 'Cluster centers',
    marker = dict(
    size=12,
    color = ('rgb(122, 296, 167)')
    )
```

```
)

layout = go.Layout(
    title='1D vector',
)

traces = [trace1, trace2]

fig = go.Figure(data=traces, layout=layout)

plot(fig)
```

Take a look at the following diagram. The given code will output the following:

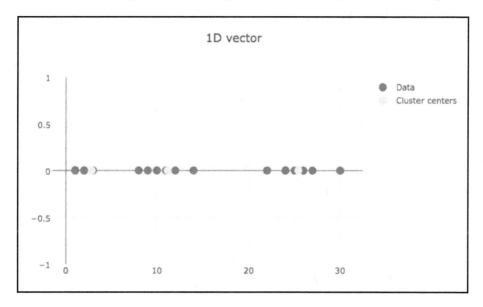

You can clearly see that each element can be assigned one of the cluster centers.

Modifying our algorithm

Now you have understood the internal of k-means on a single variable, you can extend this implementation to multiple variables and apply it to a more realistic dataset.

The dataset to be used in this section is from the *UCI Machine Learning Repository* (`https://archive.ics.uci.edu/ml/datasets/wholesale+customers`), and it includes the client information of wholesales distributor. There 440 customers with eight features. In the following list, first six features are related to annual spending for corresponding products, seventh feature shows the channel that this product is bought and the eighth feature shows the region:

- FRESH
- MILK
- GROCERY
- FROZEN
- DETERGENTS_PAPER
- DELICATESSEN
- CHANNEL
- REGION

First download the dataset and read the it as a `numpy` array:

```
from numpy import genfromtxt
wholesales_data = genfromtxt('Wholesale customers data.csv', delimiter=',',
skip_header=1)
```

You can have a quick look at the data. Here it is:

```
print(wholesales_data[:5])
[[2.0000e+00 3.0000e+00 1.2669e+04 9.6560e+03 7.5610e+03 2.1400e+02
  2.6740e+03 1.3380e+03]
 [2.0000e+00 3.0000e+00 7.0570e+03 9.8100e+03 9.5680e+03 1.7620e+03
  3.2930e+03 1.7760e+03]
 [2.0000e+00 3.0000e+00 6.3530e+03 8.8080e+03 7.6840e+03 2.4050e+03
  3.5160e+03 7.8440e+03]
 [1.0000e+00 3.0000e+00 1.3265e+04 1.1960e+03 4.2210e+03 6.4040e+03
  5.0700e+02 1.7880e+03]
 [2.0000e+00 3.0000e+00 2.2615e+04 5.4100e+03 7.1980e+03 3.9150e+03
  1.7770e+03 5.1850e+03]]
```

Checking `shape` will show you the number of rows and variables, as shown here:

```
wholesales_data.shape
# (440, 8)
```

This dataset has `440` records with `8` features each.

It's a good idea to normalize your dataset, by using this code:

```
wholesales_data_norm = wholesales_data / np.linalg.norm(wholesales_data)

print(wholesales_data_norm[:5])
[[ 1.          0.          0.30168043  1.06571214  0.32995207 -0.46657183
   0.50678671  0.2638102 ]
 [ 1.          0.         -0.1048095   1.09293385  0.56599336  0.08392603
   0.67567015  0.5740085 ]
 [ 1.          0.         -0.15580183  0.91581599  0.34441798  0.3125889
   0.73651183  4.87145892]
 [ 0.          0.          0.34485007 -0.42971408 -0.06286202  1.73470839
  -0.08444172  0.58250708]
 [ 1.          0.          1.02209184  0.3151708   0.28726     0.84957326
   0.26205579  2.98831445]]
```

You can read the dataset into the `pandas.DataFrame`, by using this code:

```
import pandas as pd

df = pd.DataFrame(wholesales_data_norm,
columns=['Channel',
'Region',
'Fresh',
'Milk',
'Grocery',
'Frozen',
'Detergents_Paper',
'Delicatessen'])

df.head(10)
```

	Channel	Region	Fresh	Milk	Grocery	Frozen	Detergents_Paper	Delicassen
0	1.0	0.0	0.301680	1.065712	0.329952	-0.466572	0.506787	0.263810
1	1.0	0.0	-0.104810	1.092934	0.565993	0.083926	0.675670	0.574008
2	1.0	0.0	-0.155802	0.915816	0.344418	0.312589	0.736512	4.871459
3	0.0	0.0	0.344850	-0.429714	-0.062862	1.734708	-0.084442	0.582507
4	1.0	0.0	1.022092	0.315171	0.287260	0.849573	0.262056	2.988314
5	1.0	0.0	0.065841	0.818772	0.043574	-0.305832	0.266967	0.343839
6	1.0	0.0	0.262350	-0.075655	0.261033	-0.371977	0.633927	-0.297805
7	1.0	0.0	-0.067000	0.234920	0.549293	0.050853	0.683309	1.133499
8	0.0	0.0	-0.184050	0.003712	0.168945	-0.391536	0.245413	-0.152620
9	1.0	0.0	-0.180936	1.319722	1.661286	-0.130512	1.803015	0.802054

Let's create a `scatterplot` matrix to have a closer look at the dataset. Take a look at the code:

```
fig = ff.create_scatterplotmatrix(df, diag='histogram', size=7,
height=1200, width=1200)
plot(fig)
```

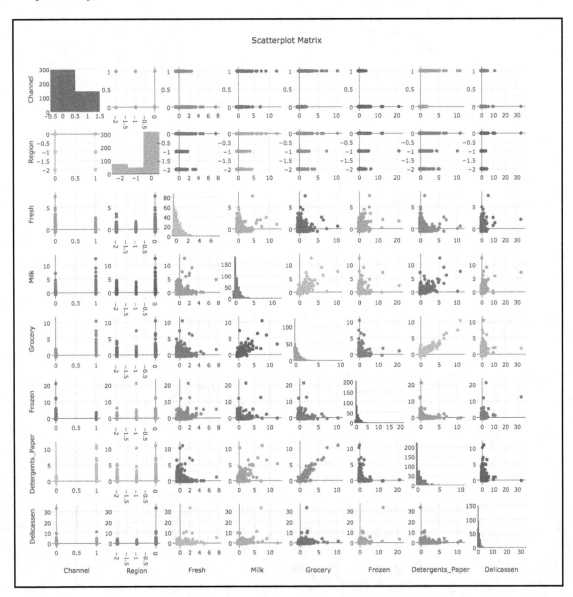

You can also check the correlations between features by running the following command:

```
df.corr()
```

This will give you a correlation table, as shown here:

	Channel	Region	Fresh	Milk	Grocery	Frozen	Detergents_Paper	Delicassen
Channel	1.000000	0.062028	-0.169172	0.460720	0.608792	-0.202046	0.636026	0.056011
Region	0.062028	1.000000	0.055287	0.032288	0.007696	-0.021044	-0.001483	0.045212
Fresh	-0.169172	0.055287	1.000000	0.100510	-0.011854	0.345881	-0.101953	0.244690
Milk	0.460720	0.032288	0.100510	1.000000	0.728335	0.123994	0.661816	0.406368
Grocery	0.608792	0.007696	-0.011854	0.728335	1.000000	-0.040193	0.924641	0.205497
Frozen	-0.202046	-0.021044	0.345881	0.123994	-0.040193	1.000000	-0.131525	0.390947
Detergents_Paper	0.636026	-0.001483	-0.101953	0.661816	0.924641	-0.131525	1.000000	0.069291
Delicassen	0.056011	0.045212	0.244690	0.406368	0.205497	0.390947	0.069291	1.000000

You can also use `seaborn` heatmap, as shown here:

```
import seaborn as sns; sns.set()
ax = sns.heatmap(df.corr(), annot=True)
```

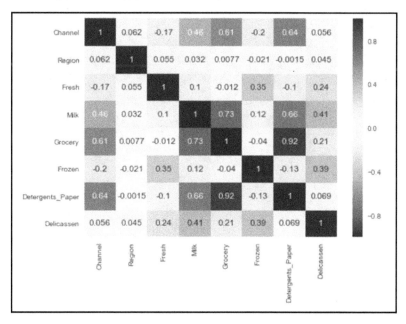

Correlations between the features

You can see that there are some strong correlations between features, such as the correlation between `Grocery` and `Detergents_Paper`.

Let's plot three features, `Grocery`, `Detergents_Paper`, and `Milk`, by using this code:

```
# Creating data for the plotly
trace1 = go.Scatter3d(
  # Extracting data based on label
  x=df['Grocery'],
  y=df['Detergents_Paper'],
  z=df['Milk'],
  mode='markers',
  marker=dict(
    size=12,
    line=dict(
    color='rgba(217, 217, 217, 0.14)',
    width=0.5
    ),
    opacity=0.8
    )
  )

# Layout settings
layout = go.Layout(
  scene = dict(
    xaxis = dict(
      title= 'Grocery'),
    yaxis = dict(
      title= 'Detergents_Paper'),
    zaxis = dict(
      title= 'Milk'),),
)

data = [trace1]

fig = dict(data=data, layout=layout)

plot(fig)
```

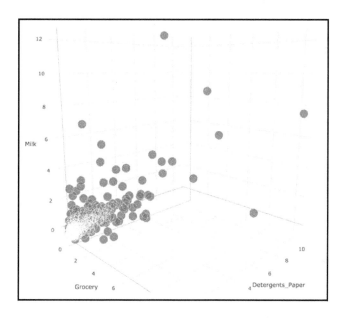

You will now go ahead and extend the k-means algorithm you have implemented for higher dimensions. First, you can drop `Channel` and `Region` from the dataset, as shown here:

```
df = df[[col for col in df.columns if col not in ['Channel', 'Region']]]
df.head(10)
```

	Fresh	Milk	Grocery	Frozen	Detergents_Paper	Delicassen
0	0.301680	1.065712	0.329952	-0.466572	0.506787	0.263810
1	-0.104810	1.092934	0.565993	0.083926	0.675670	0.574008
2	-0.155802	0.915816	0.344418	0.312589	0.736512	4.871459
3	0.344850	-0.429714	-0.062862	1.734708	-0.084442	0.582507
4	1.022092	0.315171	0.287260	0.849573	0.262056	2.988314
5	0.065841	0.818772	0.043574	-0.305832	0.266967	0.343839
6	0.262350	-0.075655	0.261033	-0.371977	0.633927	-0.297805
7	-0.067000	0.234920	0.549293	0.050853	0.683309	1.133499
8	-0.184050	0.003712	0.168945	-0.391536	0.245413	-0.152620
9	-0.180936	1.319722	1.661286	-0.130512	1.803015	0.802054

In terms of implementation, you can also use `np.linalg.norm` for calculating distances, and it's really up to you what kind of distance function you use. Another alternative is `distance.euclidean` from `scipy.spatial`, as shown here:

```
def Kmeans_nD(X, n_clusters, random_seed=442):

    # Randomly choose random indexes as cluster centers
    rng = np.random.RandomState(random_seed)
    i = rng.permutation(X.shape[0])[:n_clusters]
    c_centers = X[i]

    # Calculate distances between each point and cluster centers
    deltas = np.array([[np.linalg.norm(i - c) for c in c_centers] for i in
X])

    # Get labels for each point
    labels = deltas.argmin(1)

    while True:
      # Calculate mean of each cluster
      new_c_centers = np.array([X[np.where(deltas.argmin(1) ==
i)[0]].mean(axis=0) for i in range(n_clusters)])

      # Calculate distances again
      deltas = np.array([[np.linalg.norm(i - c) for c in new_c_centers] for i
in X])

      # Get new labels for each point
      labels = deltas.argmin(1)

      # If there's no change in centers, exit
      if np.array_equal(c_centers, new_c_centers):
        break
      c_centers = new_c_centers

    return c_centers, labels
```

`Grocery` and `Detergents_Paper` will be used in clustering, and `k` will be set as 3. Normally, you should use visual inspection or an elbow method to decide `k`, as shown here:

```
centers, labels = Kmeans_nD(df[['Grocery', 'Detergents_Paper']].values, 3)
```

Now, you can add one more column to your dataset, by using the following:

```
df['labels'] = labels
```

You can visualize the result first to see whether the results make sense, by using this code:

```
# Creating data for the plotly
trace1 = go.Scatter(
    # Extracting data based on label
    x=df[df['labels'] == 0]['Grocery'],
    y=df[df['labels'] == 0]['Detergents_Paper'],
    mode='markers',
    name='clust_1',
    marker=dict(
        size=12,
        line=dict(
        color='rgba(217, 217, 217, 0.14)',
        width=0.5
        ),
        opacity=0.8
    )
)

trace2 = go.Scatter(
    # Extracting data based on label
    x=df[df['labels'] == 1]['Grocery'],
    y=df[df['labels'] == 1]['Detergents_Paper'],
    mode='markers',
    name='clust_2',
    marker=dict(
        color='rgb(#3742fa)',
        size=12,
        symbol='circle',
        line=dict(
        color='rgb(204, 204, 204)',
        width=1
        ),
        opacity=0.9
    )
)

trace3 = go.Scatter(
    # Extracting data based on label
    x=df[df['labels'] == 2]['Grocery'],
    y=df[df['labels'] == 2]['Detergents_Paper'],
    mode='markers',
    name='clust_3',
    marker=dict(
        color='rgb(#ff4757)',
        size=12,
        symbol='circle',
        line=dict(
```

```
            color='rgb(104, 74, 114)',
            width=1
            ),
            opacity=0.9
        )
    )

data = [trace1, trace2, trace3]

# Layout settings
layout = go.Layout(
    scene = dict(
    xaxis = dict(
    title= 'Grocery'),
    yaxis = dict(title= 'Detergents_Paper'),
    )
)

fig = go.Figure(data=data, layout=layout)

plot(fig)
```

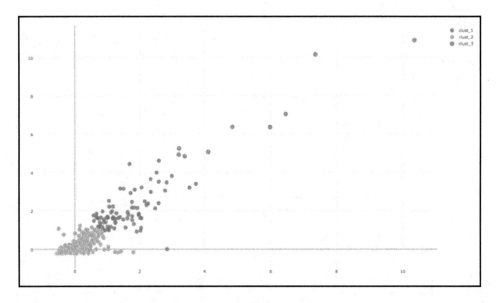

Plotting the clusters

At first glance, the clusters may look reasonable, and will ultimately depend on the interpretation that will be supported by domain knowledge.

You can easily look at the average spend for each feature per cluster, by using this code:

```
df.groupby('labels').mean()
```

labels	Fresh	Milk	Grocery	Frozen	Detergents_Paper	Delicassen
0	-0.070577	1.316079	1.558293	0.037966	1.963503	0.665974
1	0.320582	0.026393	-0.036936	0.671540	0.068682	0.302569
2	0.635915	5.588699	5.432275	0.560929	6.780301	1.526479

You can see from this simple clustering that the third cluster has the highest spenders for `Milk`, `Grocery`, and `Detergents_Paper`. The second cluster has low spenders, and the first cluster inclines toward `Milk`, `Grocery`, and `Detergents_Paper`, hence k=2 might work too.

Summary

In this chapter, you have learned the basics of unsupervised learning and using the k-means algorithm for clustering.

There are many clustering algorithms that show different behavior. Visualization is key when it comes to unsupervised learning algorithms, and you have seen a couple of different ways to visualize and inspect your dataset.

In the next chapter, you will learn other libraries which are commonly used with NumPy such as SciPy, Pandas and scikit-learn. These are all important libraries in the practitioner's toolkit, and they complement one another. You will find yourself using these libraries together with NumPy, as each will make certain tasks easier; hence, it's important to know more about the Python data science stack.

6
NumPy, SciPy, Pandas, and Scikit-Learn

By now, you should be able to write small implementations with NumPy. Throughout the chapters, we aim to provide examples using other libraries as well and in this chapter, we should step back and look at the surrounding libraries that you can use along with NumPy for your projects.

We will be covering how other Python libraries complement NumPy in this chapter. We will be looking at the following topics:

- NumPy and SciPy
- NumPy and pandas
- SciPy and scikit-learn

NumPy and SciPy

Until now, you have seen numerous examples of NumPy usage and only a few of SciPy. NumPy has array data type, which allows you to perform various array operations, such as sorting and reshaping.

 NumPy has some numerical algorithms that can be used for tasks such as calculating norms, eigenvalues, and eigenvectors. However, if numerical algorithms are your focus, you should ideally use SciPy, as it includes a more comprehensive algorithm set, as well as the latest versions of the algorithms. SciPy has a lot of useful subpackages for certain kinds of analysis.

The following list will give you an overall idea of the subpackages:

- `Cluster`: This subpackage includes clustering algorithms. It has two submodules, `vq` and `hierarchy`. The `vq` module provides functions for *k*-means clustering. The *hierarchy* module includes functions for hierarchical clustering.
- `Fftpack`: This subpackage includes functions and algorithms for fast Fourier transforms, as well as differential and pseudodifferential operators.
- `Interpolate`: This subpackage provides functions for univariate and multivariate interpolation: 1D and 2D splines.
- `Linalg`: This subpackage provides functions and algorithms for linear algebra, such as `matrix` operations and functions, eigenvalue and -vector calculations, matrix decompositions, matrix equation solvers, and special matrices.
- `Ndimage`: This subpackage provides functions and algorithms for multidimensional image processing, such as filters, interpolation, measurements, and morphology.
- `Optimize`: This subpackage provides functions and algorithms for function-local and -global optimization, function fitting, root finding, and linear programming.
- `Signal`: This subpackage provides functions and algorithms for signal processing, such as convolution, *b*-splines, filtering, continuous and discrete-time linear systems, waveforms, wavelets, and spectral analysis.
- `Stats`: This subpackage provides probability distributions, such as continuous, multivariate, and discrete distributions, and statistical functions that can find the mean, mode, variance, skewness, kurtosis, and correlation coefficients.

Let's see one of these subpackages in action. The following code shows a `cluster` package, used in cluster analysis:

```
Scipy.cluster

%matplotlib inline
import matplotlib.pyplot as plt

# Import ndimage to read the image
from scipy import ndimage

# Import cluster for clustering algorithms
from scipy import cluster

# Read the image
image = ndimage.imread("cluster_test_image.jpg")
# Image is 1000x1000 pixels and it has 3 channels.
image.shape
```

```
(1000, 1000, 3)
```

This will give you the following output:

```
array([[[30, 30, 30],
        [16, 16, 16],
        [14, 14, 14],
        ...,
        [14, 14, 14],
        [16, 16, 16],
        [29, 29, 29]],

       [[13, 13, 13],
        [ 0,  0,  0],
        [ 0,  0,  0],
        ...,
        [ 0,  0,  0],
        [ 0,  0,  0],
        [12, 12, 12]],

       [[16, 16, 16],
        [ 3,  3,  3],
        [ 1,  1,  1],
        ...,
        [ 0,  0,  0],
        [ 2,  2,  2],
        [16, 16, 16]],

       ...,

       [[17, 17, 17],
        [ 3,  3,  3],
        [ 1,  1,  1],
        ...,
        [34, 26, 39],
        [27, 21, 33],
        [59, 55, 69]],

       [[15, 15, 15],
        [ 2,  2,  2],
        [ 0,  0,  0],
        ...,
        [37, 31, 43],
        [34, 28, 42],
        [60, 56, 71]],

       [[33, 33, 33],
        [20, 20, 20],
```

```
[17, 17, 17],
...,
[55, 49, 63],
[47, 43, 57],
[65, 61, 76]]], dtype=uint8)
```

Here, you can see the plot:

```
plt.figure(figsize = (15,8))
plt.imshow(image)
```

You get the following plot from the preceding code block:

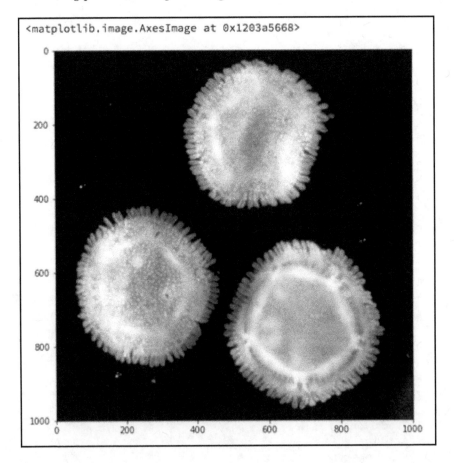

Convert image array to a 2-dimensional dataset, by using this code:

```
x, y, z = image.shape
image_2d = image.reshape(x*y, z).astype(float)
image_2d.shape

(1000000, 3)

image_2d

array([[30., 30., 30.],
       [16., 16., 16.],
       [14., 14., 14.],
       ...,
       [55., 49., 63.],
       [47., 43., 57.],
       [65., 61., 76.]])


# kmeans will return cluster centers and the distortion
cluster_centers, distortion = cluster.vq.kmeans(image_2d, k_or_guess=2)

print(cluster_centers, distortion)

[[179.28653454 179.30176248 179.44142117]
 [  3.75308484   3.83491111   4.49236356]] 26.87835069294931

image_2d_labeled = image_2d.copy()

labels = []

from scipy.spatial.distance import euclidean
import numpy as np

for i in range(image_2d.shape[0]):
    distances = [euclidean(image_2d[i], center) for center in
cluster_centers]
    labels.append(np.argmin(distances))

plt.figure(figsize = (15,8))
plt.imshow(cluster_centers[labels].reshape(x, y, z))
```

You get the following output from the preceding code:

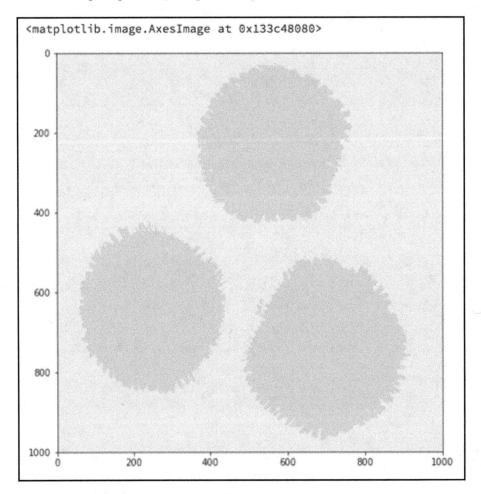

```
<matplotlib.image.AxesImage at 0x133c48080>
```

Linear regression with SciPy and NumPy

You have already seen how to code a linear regression algorithm from scratch with NumPy. The Scipy.stats module has a linregress function to calculate the slope, intercept, correlation coefficient (*r*-value), two-sided *p*-value, and standard error of the estimate, as shown here:

```
from sklearn import datasets
%matplotlib inline
```

```
import matplotlib.pyplot as plt

# Boston House Prices dataset
boston = datasets.load_boston()
x = boston.data
y = boston.target

boston.feature_names

array(['CRIM', 'ZN', 'INDUS', 'CHAS', 'NOX', 'RM', 'AGE', 'DIS', 'RAD',
       'TAX', 'PTRATIO', 'B', 'LSTAT'], dtype='<U7')

x.shape
(506, 13)

y.shape
(506,)

# We will consider "lower status of population" as independent variable for
its importance
lstat = x[0:,-1]
lstat.shape
(506,)

from scipy import stats

slope, intercept, r_value, p_value, std_err = stats.linregress(lstat, y)

print(slope, intercept, r_value, p_value, std_err)

-0.9500493537579909 34.55384087938311 -0.737662726174015
5.081103394387796e-88 0.03873341621263942

print("r-squared:", r_value**2)
r-squared: 0.5441462975864798

plt.plot(lstat, y, 'o', label='original data')
plt.plot(lstat, intercept + slope*lstat, 'r', label='fitted line')
plt.legend()
plt.show()
```

We get the following plot from the output of the preceding code, as shown in this diagram:

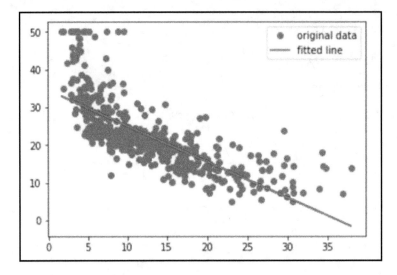

You can also have a look at the relationship between *average number of rooms* and the house prices. Following code block prints out the performance metric:

```
rm = x[0:,5]

slope, intercept, r_value, p_value, std_err = stats.linregress(rm, y)

print(slope, intercept, r_value, p_value, std_err)

print("r-squared:", r_value**2)

# 9.102108981180308 -34.670620776438554 0.6953599470715394
2.48722887100781e-74 0.4190265601213402
# r-squared: 0.483525455991334
```

The following code block plots the fitted line:

```
plt.plot(rm, y, 'o', label='original data')
plt.plot(rm, intercept + slope*rm, 'r', label='fitted line')
plt.legend()
plt.show()
```

We get the following output from the preceding code, as shown in this diagram:

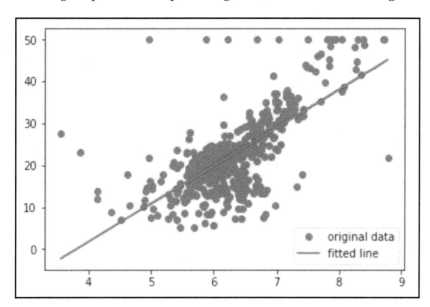

NumPy and pandas

When you think about it, NumPy is a fairly low-level array-manipulation library, and the majority of other Python libraries are written on top of it.

One of these libraries is `pandas`, which is a high-level data-manipulation library. When you are exploring a dataset, you usually perform operations such as calculating descriptive statistics, grouping by a certain characteristic, and merging. The `pandas` library has many friendly functions to perform these various useful operations.

Let's use a diabetes dataset in this example. The diabetes dataset in `sklearn.datasets` is standardized with a zero mean and unit L2 norm.

The dataset contains 442 records with 10 features: age, sex, body mass index, average blood pressure, and six blood serum measurements.

The target represents the disease progression after these baseline measures are taken. You can look at the data description at `https://www4.stat.ncsu.edu/~boos/var.select/diabetes.html` and a related paper at `http://web.stanford.edu/~hastie/Papers/LARS/LeastAngle_2002.pdf`.

We start with our operation, as follows:

```
import pandas as pd
from sklearn import datasets

%matplotlib inline
import matplotlib.pyplot as plt
import seaborn as sns

diabetes = datasets.load_diabetes()

df = pd.DataFrame(diabetes.data, columns=diabetes.feature_names)

diabetes.feature_names
['age', 'sex', 'bmi', 'bp', 's1', 's2', 's3', 's4', 's5', 's6']

df.head(10)
```

We get the following output from the preceding code, as shown in this table:

	age	sex	bmi	bp	s1	s2	s3	s4	s5	s6	Target
0	0.038076	0.050680	0.061696	0.021872	-0.044223	-0.034821	-0.043401	-0.002592	0.019908	-0.017646	151.0
1	-0.001882	-0.044642	-0.051474	-0.026328	-0.008449	-0.019163	0.074412	-0.039493	-0.068330	-0.092204	75.0
2	0.085299	0.050680	0.044451	-0.005671	-0.045599	-0.034194	-0.032356	-0.002592	0.002864	-0.025930	141.0
3	-0.089063	-0.044642	-0.011595	-0.036656	0.012191	0.024991	-0.036038	0.034309	0.022692	-0.009362	206.0
4	0.005383	-0.044642	-0.036385	0.021872	0.003935	0.015596	0.008142	-0.002592	-0.031991	-0.046641	135.0
5	-0.092695	-0.044642	-0.040696	-0.019442	-0.068991	-0.079288	0.041277	-0.076395	-0.041180	-0.096346	97.0
6	-0.045472	0.050680	-0.047163	-0.015999	-0.040096	-0.024800	0.000779	-0.039493	-0.062913	-0.038357	138.0
7	0.063504	0.050680	-0.001895	0.066630	0.090620	0.108914	0.022869	0.017703	-0.035817	0.003064	63.0
8	0.041708	0.050680	0.061696	-0.040099	-0.013953	0.006202	-0.028674	-0.002592	-0.014956	0.011349	110.0
9	-0.070900	-0.044642	0.039062	-0.033214	-0.012577	-0.034508	-0.024993	-0.002592	0.067736	-0.013504	310.0

This code shows you how to create a target column in DataFrame:

```
df['Target'] = diabetes.target
df.head(10)
```

	age	sex	bmi	bp	s1	s2	s3	s4	s5	s6	Target
0	0.038076	0.050680	0.061696	0.021872	-0.044223	-0.034821	-0.043401	-0.002592	0.019908	-0.017646	151.0
1	-0.001882	-0.044642	-0.051474	-0.026328	-0.008449	-0.019163	0.074412	-0.039493	-0.068330	-0.092204	75.0
2	0.085299	0.050680	0.044451	-0.005671	-0.045599	-0.034194	-0.032356	-0.002592	0.002864	-0.025930	141.0
3	-0.089063	-0.044642	-0.011595	-0.036656	0.012191	0.024991	-0.036038	0.034309	0.022692	-0.009362	206.0
4	0.005383	-0.044642	-0.036385	0.021872	0.003935	0.015596	0.008142	-0.002592	-0.031991	-0.046641	135.0
5	-0.092695	-0.044642	-0.040696	-0.019442	-0.068991	-0.079288	0.041277	-0.076395	-0.041180	-0.096346	97.0
6	-0.045472	0.050680	-0.047163	-0.015999	-0.040096	-0.024800	0.000779	-0.039493	-0.062913	-0.038357	138.0
7	0.063504	0.050680	-0.001895	0.066630	0.090620	0.108914	0.022869	0.017703	-0.035817	0.003064	63.0
8	0.041708	0.050680	0.061696	-0.040099	-0.013953	0.006202	-0.028674	-0.002592	-0.014956	0.011349	110.0
9	-0.070900	-0.044642	0.039062	-0.033214	-0.012577	-0.034508	-0.024993	-0.002592	0.067736	-0.013504	310.0

Pandas helps us to work with tabular data easily and supports our analysis with various helper methods and visualizations. Take a look at the code:

```
# Descriptive statistics
df.describe()
```

We get the following output from the preceding code, as shown in this table:

	age	sex	bmi	bp	s1	s2	s3	s4	s5	s6	Target
count	4.420000e+02	4.420000e+02	4.420000e+02	4.420000e+02	4.420000e+02	4.420000e+02	4.420000e+02	4.420000e+02	4.420000e+02	4.420000e+02	442.000000
mean	-3.634285e-16	1.308343e-16	-8.045349e-16	1.281655e-16	-8.835316e-17	1.327024e-16	-4.574646e-16	3.777301e-16	-3.830854e-16	-3.412882e-16	152.133484
std	4.761905e-02	4.761905e-02	4.761905e-02	4.761905e-02	4.761905e-02	4.761905e-02	4.761905e-02	4.761905e-02	4.761905e-02	4.761905e-02	77.093005
min	-1.072256e-01	-4.464164e-02	-9.027530e-02	-1.123996e-01	-1.267807e-01	-1.156131e-01	-1.023071e-01	-7.639450e-02	-1.260974e-01	-1.377672e-01	25.000000
25%	-3.729927e-02	-4.464164e-02	-3.422907e-02	-3.665645e-02	-3.424784e-02	-3.035840e-02	-3.511716e-02	-3.949338e-02	-3.324879e-02	-3.317903e-02	87.000000
50%	5.383060e-03	-4.464164e-02	-7.283766e-03	-5.670611e-03	-4.320866e-03	-3.819065e-03	-6.584468e-03	-2.592262e-03	-1.947634e-03	-1.077698e-03	140.500000
75%	3.807591e-02	5.068012e-02	3.124802e-02	3.564384e-02	2.835801e-02	2.984439e-02	2.931150e-02	3.430886e-02	3.243323e-02	2.791705e-02	211.500000
max	1.107267e-01	5.068012e-02	1.705552e-01	1.320442e-01	1.539137e-01	1.987880e-01	1.811791e-01	1.852344e-01	1.335990e-01	1.356118e-01	346.000000

Let's see how target is distributed, by using this line of code:

```
plt.hist(df['Target'])
```

The following diagram shows the output from the preceding line:

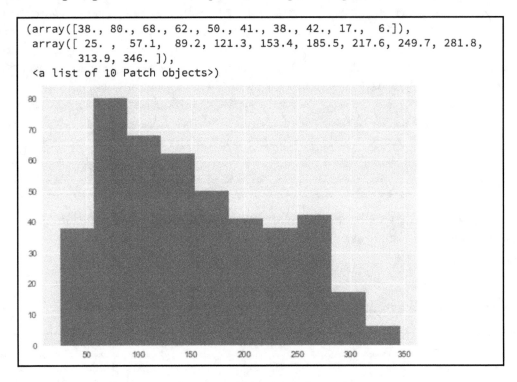

```
(array([38., 80., 68., 62., 50., 41., 38., 42., 17.,  6.]),
 array([ 25. ,  57.1,  89.2, 121.3, 153.4, 185.5, 217.6, 249.7, 281.8,
        313.9, 346. ]),
 <a list of 10 Patch objects>)
```

You can see that the target variable is skewed to the right. Take a look at this code:

```
# Since 'sex' is categorical, excluding it from numerical columns
numeric_cols = [col for col in df.columns if col != 'sex']

numeric_cols
# ['age', 'bmi', 'bp', 's1', 's2', 's3', 's4', 's5', 's6', 'Target']

# You can have a look at variable distributions individually, but there's a
better way
df[numeric_cols].hist(figsize=(20, 20), bins=30, xlabelsize=12,
ylabelsize=12)

# You can also choose create dataframes for numerical and categorical
variables
```

Output from the preceding code block:

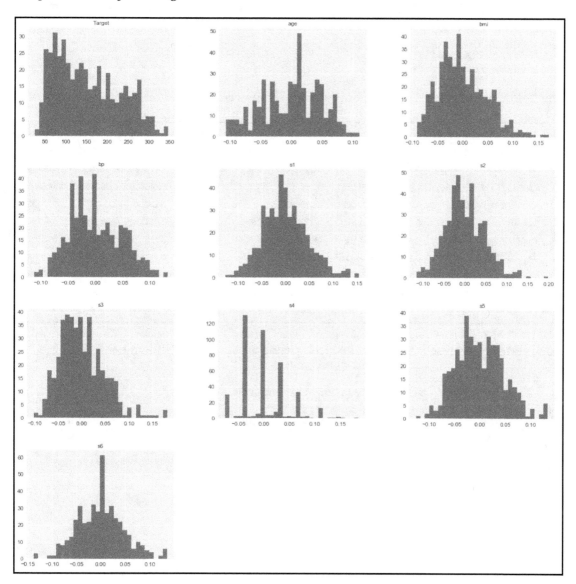

Feature distributions

You can examine the distribution of several of the features and decide which of them looks similar. For this example, features **s1**, **s2**, and **s6** seem to have a similar distribution, as you can see from this code:

```
# corr method will give you the correlation between features
df[numeric_cols].corr()
```

The following diagram shows the output from the preceding line:

	age	bmi	bp	s1	s2	s3	s4	s5	s6	Target
age	1.000000	0.185085	0.335427	0.260061	0.219243	-0.075181	0.203841	0.270777	0.301731	0.187889
bmi	0.185085	1.000000	0.395415	0.249777	0.261170	-0.366811	0.413807	0.446159	0.388680	0.586450
bp	0.335427	0.395415	1.000000	0.242470	0.185558	-0.178761	0.257653	0.393478	0.390429	0.441484
s1	0.260061	0.249777	0.242470	1.000000	0.896663	0.051519	0.542207	0.515501	0.325717	0.212022
s2	0.219243	0.261170	0.185558	0.896663	1.000000	-0.196455	0.659817	0.318353	0.290600	0.174054
s3	-0.075181	-0.366811	-0.178761	0.051519	-0.196455	1.000000	-0.738493	-0.398577	-0.273697	-0.394789
s4	0.203841	0.413807	0.257653	0.542207	0.659817	-0.738493	1.000000	0.617857	0.417212	0.430453
s5	0.270777	0.446159	0.393478	0.515501	0.318353	-0.398577	0.617857	1.000000	0.464670	0.565883
s6	0.301731	0.388680	0.390429	0.325717	0.290600	-0.273697	0.417212	0.464670	1.000000	0.382483
Target	0.187889	0.586450	0.441484	0.212022	0.174054	-0.394789	0.430453	0.565883	0.382483	1.000000

You can better represent this relationship by using a `heatmap`, as shown here:

```
plt.figure(figsize=(15, 15))
sns.heatmap(df[numeric_cols].corr(), annot=True)
```

The following diagram is a heatmap that was generated by the preceding code block:

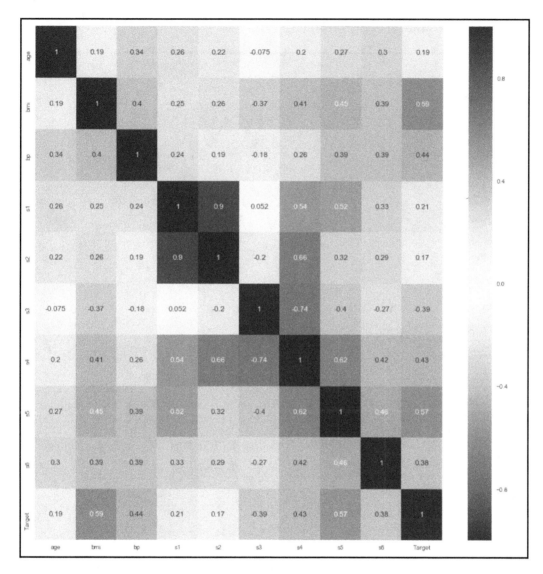

Correlations heatmap

You can also filter correlations in the following way:

```
plt.figure(figsize=(18, 15))
sns.heatmap(df[numeric_cols].corr()
            [(df[numeric_cols].corr() >= 0.3) & (df[numeric_cols].corr() <=
0.5)],
            annot=True)
```

This diagram shows the filtered correlations:

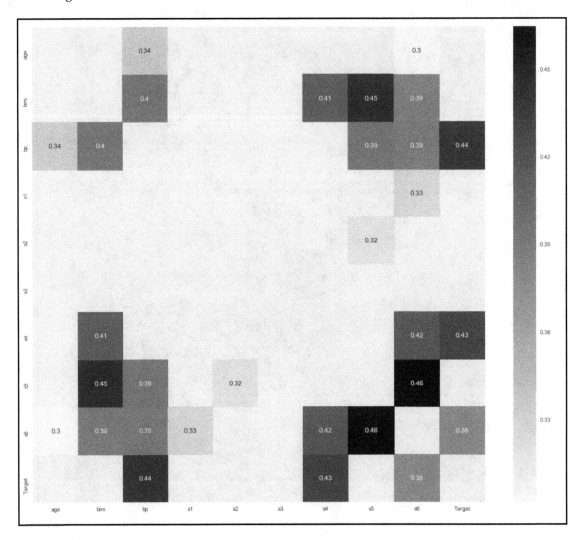

Filtered correlations heatmap

There are other useful visualizations that you can use to examine statistical relationships, as shown here:

```
fig, ax = plt.subplots(3, 3, figsize = (18, 12))
for i, ax in enumerate(fig.axes):
    if i < 9:
        sns.regplot(x=df[numeric_cols[i]],y='Target', data=df, ax=ax)
```

This diagram shows the following output from the preceding code:

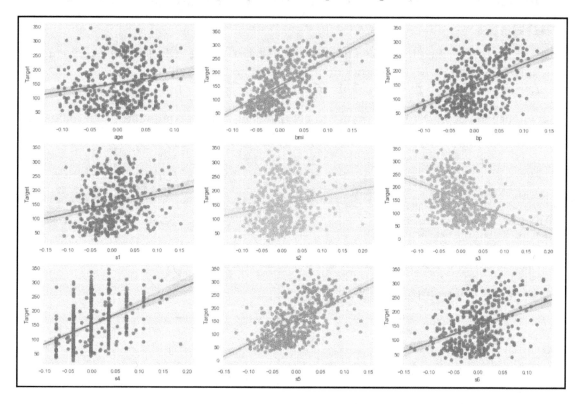

Regression Plots

You can see that using `pandas` makes exploratory data analysis relatively simpler. Using `pandas`, you can inspect features and their relationships.

Quantitative modeling with stock prices using pandas

`pandas` was first written to be used in financial datasets, and it includes many convenient functions for dealing with time series data. In this section, you will see how you can handle stock price series, using the `pandas` library.

You will use the `quandl` Python library to get financial data for companies. Take a look at this code:

```
import quandl msft = quandl.get('WIKI/MSFT')

msft.columns
# Index(['Open', 'High', 'Low', 'Close', 'Volume', 'Ex-Dividend', 'Split
Ratio', 'Adj. Open', 'Adj. High', 'Adj. Low', 'Adj. Close', 'Adj. Volume'],
dtype='object')

msft.tail()
```

This table shows the output of `msft.tail()`:

Date	Open	High	Low	Close	Volume	Ex-Dividend	Split Ratio	Adj. Open	Adj. High	Adj. Low	Adj. Close	Adj. Volume
2018-03-21	92.930	94.050	92.21	92.48	23753263.0	0.0	1.0	92.930	94.050	92.21	92.48	23753263.0
2018-03-22	91.265	91.750	89.66	89.79	37578166.0	0.0	1.0	91.265	91.750	89.66	89.79	37578166.0
2018-03-23	89.500	90.460	87.08	87.18	42159397.0	0.0	1.0	89.500	90.460	87.08	87.18	42159397.0
2018-03-26	90.610	94.000	90.40	93.78	55031149.0	0.0	1.0	90.610	94.000	90.40	93.78	55031149.0
2018-03-27	94.940	95.139	88.51	89.47	53704562.0	0.0	1.0	94.940	95.139	88.51	89.47	53704562.0

Let's customize our plot with the following settings import:

```
matplotlib.font_manager as font_manager font_path =
'/Library/Fonts/Cochin.ttc' font_prop =
font_manager.FontProperties(fname=font_path, size=24) axis_font =
{'fontname':'Arial', 'size':'18'} title_font = {'fontname':'Arial',
'size':'22', 'color':'black', 'weight':'normal',
'verticalalignment':'bottom'} plt.figure(figsize=(10, 8))
plt.plot(msft['Adj. Close'], label='Adj. Close') plt.xticks(fontsize=22)
plt.yticks(fontsize=22) plt.xlabel("Date", **axis_font) plt.ylabel("Adj.
Close", **axis_font) plt.title("MSFT", **title_font) plt.legend(loc='upper
left', prop=font_prop, numpoints=1) plt.show()
```

This diagram shows the plot from the preceding settings:

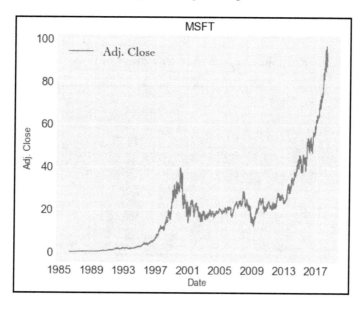

You can use this code to calculate daily change:

```
msft['Daily Pct. Change'] = (msft['Adj. Close'] - msft['Adj. Open']) /
msft['Adj. Open']

msft.tail(10)
```

This diagram shows the output of `msft.tail(10)`:

Date	Open	High	Low	Close	Volume	Ex-Dividend	Split Ratio	Adj. Open	Adj. High	Adj. Low	Adj. Close	Adj. Volume	Daily Pct. Change
2018-03-14	95.120	95.410	93.50	93.85	31576898.0	0.0	1.0	95.120	95.410	93.50	93.85	31576898.0	-0.013352
2018-03-15	93.530	94.580	92.83	94.18	26279014.0	0.0	1.0	93.530	94.580	92.83	94.18	26279014.0	0.006950
2018-03-16	94.680	95.380	93.92	94.60	47329521.0	0.0	1.0	94.680	95.380	93.92	94.60	47329521.0	-0.000845
2018-03-19	93.740	93.900	92.11	92.89	31752589.0	0.0	1.0	93.740	93.900	92.11	92.89	31752589.0	-0.009068
2018-03-20	93.050	93.770	93.00	93.13	21787780.0	0.0	1.0	93.050	93.770	93.00	93.13	21787780.0	0.000860
2018-03-21	92.930	94.050	92.21	92.48	23753263.0	0.0	1.0	92.930	94.050	92.21	92.48	23753263.0	-0.004842
2018-03-22	91.265	91.750	89.66	89.79	37578166.0	0.0	1.0	91.265	91.750	89.66	89.79	37578166.0	-0.016162
2018-03-23	89.500	90.460	87.08	87.18	42159397.0	0.0	1.0	89.500	90.460	87.08	87.18	42159397.0	-0.025922
2018-03-26	90.610	94.000	90.40	93.78	55031149.0	0.0	1.0	90.610	94.000	90.40	93.78	55031149.0	0.034985
2018-03-27	94.940	95.139	88.51	89.47	53704562.0	0.0	1.0	94.940	95.139	88.51	89.47	53704562.0	-0.057615

Try using this histogram of daily returns:

```
plt.figure(figsize=(22, 8))
plt.hist(msft['Daily Pct. Change'], bins=100)
```

It will give you the following plot, as shown in this diagram:

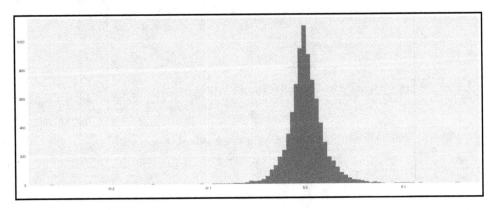

Distribution of daily returns

The returned distribution has long tails, especially on the negative side, and this is a known phenomenon in financial analysis. The risk that it creates is referred to as the tail risk, and it contradicts the assumption that the market returns follow normal distribution. This basically tells you that the probability of extreme events happening is more likely than that of a more normal distribution.

In terms of visualizations, it's helpful to make them interactive. For this, `plotly` offers a great alternative to current plotting libraries, as shown here:

```
import plotly.plotly as py
import plotly.graph_objs as go
from plotly.offline import download_plotlyjs, init_notebook_mode, plot,
iplot init_notebook_mode(connected=True)
from datetime import datetime
import pandas_datareader.data as web
import quandl

msft = quandl.get('WIKI/MSFT')
msft['Daily Pct. Change'] = (msft['Adj. Close'] - msft['Adj. Open']) /
msft['Adj. Open']

data = [go.Scatter(x=msft.index, y=msft['Adj. Close'])]

plot(data)
```

We get the following plot from the preceding code, as shown in this diagram:

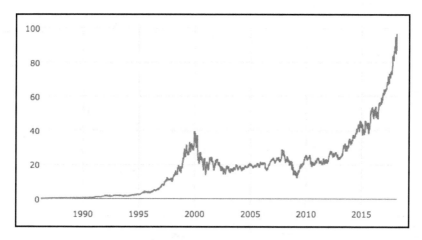

You can create an **open-high-low-close** (**OHLC**) chart, where each date has 4 different price values, which are open, high, low and close. Take a look at this code:

```
charts trace = go.Ohlc(x=msft.index, open=msft['Adj. Open'],
high=msft['Adj. High'], low=msft['Adj. Low'], close=msft['Adj. Close'])

data = [trace]

plot(data)
```

This diagram shows the plot from the preceding code:

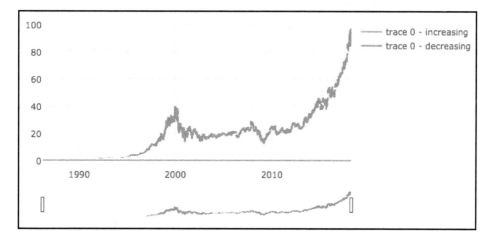

You can define custom ranges in which to inspect a specific region by selecting them on the chart. Take a look at this diagram:

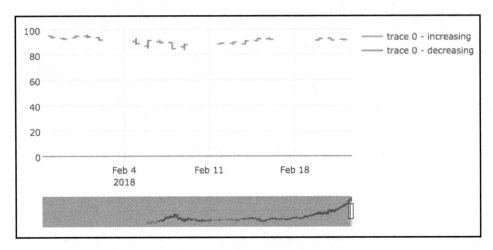

Similarly, you can create `Candlestick` charts, by using this code:

```
trace = go.Candlestick(x=msft.index, open=msft['Adj. Open'],
high=msft['Adj. High'], low=msft['Adj. Low'], close=msft['Adj. Close'])

data = [trace]

plot(data)
```

This diagram shows the output of this code:

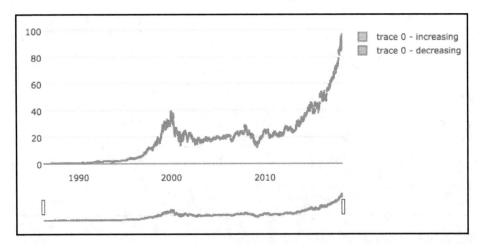

You can select specific ranges in `Candlestick` charts as well. Take a look at this diagram:

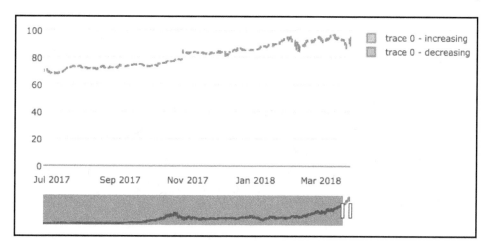

The distribution plot is as follows:

```
import plotly.figure_factory as ff

fig = ff.create_distplot([msft['Daily Pct. Change'].values], ['MSFT Daily
Returns'], show_hist=False)

plot(fig)
```

The following diagram shows the output of the preceding code:

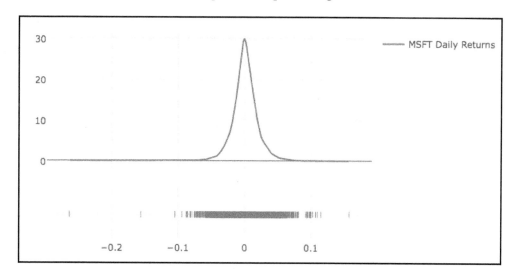

We can create three moving averages, as follows:

```
msft['200MA'] = msft['Adj. Close'].rolling(window=200).mean()
msft['100MA'] = msft['Adj. Close'].rolling(window=100).mean()
msft['50MA'] = msft['Adj. Close'].rolling(window=50).mean()

msft.tail(10)
```

This table shows the output of `msft.tail(10)`:

Date	Open	High	Low	Close	Volume	Ex-Dividend	Split Ratio	Adj. Open	Adj. High	Adj. Low	Adj. Close	Adj. Volume	Daily Pct. Change	200MA	100MA	50MA
2018-03-14	95.120	95.410	93.50	93.85	31576898.0	0.0	1.0	95.120	95.410	93.50	93.85	31576898.0	-0.013352	79.764181	87.322623	91.4226
2018-03-15	93.530	94.580	92.83	94.18	26279014.0	0.0	1.0	93.530	94.580	92.83	94.18	26279014.0	0.006950	79.888837	87.492232	91.5872
2018-03-16	94.680	95.380	93.92	94.60	47329521.0	0.0	1.0	94.680	95.380	93.92	94.60	47329521.0	-0.000845	80.013416	87.663055	91.7522
2018-03-19	93.740	93.900	92.11	92.89	31752589.0	0.0	1.0	93.740	93.900	92.11	92.89	31752589.0	-0.009068	80.132286	87.807824	91.8678
2018-03-20	93.050	93.770	93.00	93.13	21787780.0	0.0	1.0	93.050	93.770	93.00	93.13	21787780.0	0.000860	80.251028	87.954794	91.9666
2018-03-21	92.930	94.050	92.21	92.48	23753263.0	0.0	1.0	92.930	94.050	92.21	92.48	23753263.0	-0.004842	80.358327	88.094965	92.0506
2018-03-22	91.265	91.750	89.66	89.79	37578166.0	0.0	1.0	91.265	91.750	89.66	89.79	37578166.0	-0.016162	80.449602	88.210525	92.0820
2018-03-23	89.500	90.460	87.08	87.18	42159397.0	0.0	1.0	89.500	90.460	87.08	87.18	42159397.0	-0.025922	80.526639	88.298691	92.0692
2018-03-26	90.610	94.000	90.40	93.78	55031149.0	0.0	1.0	90.610	94.000	90.40	93.78	55031149.0	0.034985	80.637320	88.402612	92.1832
2018-03-27	94.940	95.139	88.51	89.47	53704562.0	0.0	1.0	94.940	95.139	88.51	89.47	53704562.0	-0.057615	80.728653	88.462637	92.1810

Based on sliced data, the last 2,000 days will be included. Take a look at this code:

```
trace_adjclose = go.Scatter( x=msft[-2000:].index, y=msft[-2000:]['Adj.
Close'], name = "Adj. Close", line = dict(color = '#000000'), opacity =
0.8)

trace_200 = go.Scatter( x=msft[-2000:].index, y=msft[-2000:]['200MA'], name
= "200MA", line = dict(color = '#FF0000'), opacity = 0.8)

trace_100 = go.Scatter( x=msft[-2000:].index, y=msft[-2000:]['100MA'], name
= "100MA", line = dict(color = '#0000FF'), opacity = 0.8)

trace_50 = go.Scatter( x=msft[-2000:].index, y=msft[-2000:]['50MA'], name =
"50MA", line = dict(color = '#FF00FF'), opacity = 0.8)

data = [trace_adjclose, trace_200, trace_100, trace_50]

layout = dict( title = "MSFT Moving Averages: 200, 100, 50 days", )

fig = dict(data=data, layout=layout)

plot(fig)
```

This diagram shows the plot from preceding code block:

Moving averages are used to monitor trends in financial markets. In this example, there are three moving averages, and each has a different period. You can set the number of days for your analysis for short, medium, and long-term trend monitoring.

When you start working with financial time series, you will quickly realize that you need aggregations based on different periods, and it's super easy to create these in `pandas`. The following snippet will aggregate records monthly by calculating the mean:

```
msft_monthly = msft.resample('M').mean()

msft_monthly.tail(10)
```

This diagram shows the output of `msft_monthly.tail(10)`:

Date	Open	High	Low	Close	Volume	Ex-Dividend	Split Ratio	Adj. Open	Adj. High	Adj. Low	Adj. Close	Adj. Volume	Daily Pct. Change	200MA	100MA	50MA
2017-06-30	70.561364	71.014800	69.835727	70.517955	2.773277e+07	0.000000	1.0	69.834054	70.282615	69.115897	69.791092	2.773277e+07	-0.000553	61.950990	65.477445	67.593782
2017-07-31	71.843250	72.412995	71.441000	72.012500	2.256239e+07	0.000000	1.0	71.102727	71.666599	70.704623	71.270232	2.256239e+07	0.002403	63.431438	66.971306	69.261458
2017-08-31	72.715652	73.196083	72.285187	72.816957	1.864639e+07	0.016957	1.0	72.183532	72.660475	71.756218	72.284124	1.864639e+07	0.001409	65.098648	68.785956	70.822208
2017-09-30	74.365500	74.786000	73.891000	74.344500	1.835672e+07	0.000000	1.0	73.990997	74.409380	73.518887	73.970103	1.835672e+07	-0.000265	66.700094	70.590506	72.419931
2017-10-31	77.889091	78.349318	77.529773	77.939545	2.002319e+07	0.000000	1.0	77.496844	77.954753	77.139335	77.547044	2.002319e+07	0.000766	68.223272	72.157040	73.922416
2017-11-30	83.620500	84.061610	83.124875	83.675500	1.980172e+07	0.021000	1.0	83.430357	83.870554	82.936030	83.485128	1.980172e+07	0.000679	70.262112	74.611141	77.520274
2017-12-31	84.836000	85.409915	84.163255	84.758500	2.237773e+07	0.000000	1.0	84.836000	85.409915	84.163255	84.758500	2.237773e+07	-0.000846	72.340131	77.291184	81.378112
2018-01-31	89.965952	90.657486	89.372143	90.074286	2.587511e+07	0.000000	1.0	89.965952	90.657486	89.372143	90.074286	2.587511e+07	0.001250	74.734287	80.352415	85.030938
2018-02-28	91.392105	92.764974	90.055832	91.413158	3.633093e+07	0.000000	1.0	91.392105	92.764974	90.055832	91.413158	3.633093e+07	0.000513	77.324649	83.868860	88.237253
2018-03-31	93.570263	94.471032	92.227684	93.169474	3.339462e+07	0.000000	1.0	93.570263	94.471032	92.227684	93.169474	3.339462e+07	-0.004179	79.714769	87.235899	91.248411

Here is a simple time-series plot:

```
data = [go.Scatter(x=msft_monthly[-24:].index, y = msft_monthly[-24:]['Adj.
Close'])]

plot(data)
```

This will give you the following plot, as shown in this diagram:

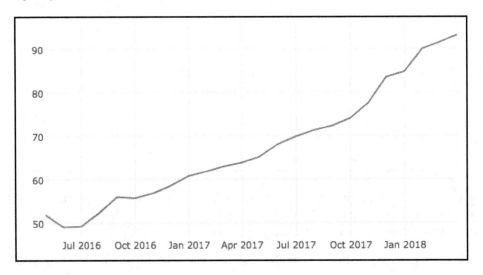

When you are inspecting a relationship between features, you can use a correlation matrix, which we have seen in previous examples. In time series, practitioners are interested in autocorrelation, which shows the correlation of the time series with itself. For example, you ideally expect to have periodic peaks in the time series that show where you have seasonality. Let's see whether the daily percentage change has any significant peaks, by using this code:

```
plt.figure(figsize=(22, 14))
pd.plotting.autocorrelation_plot(msft_monthly['Daily Pct. Change'])
```

We get the following plot from the preceding code, as shown in this diagram:

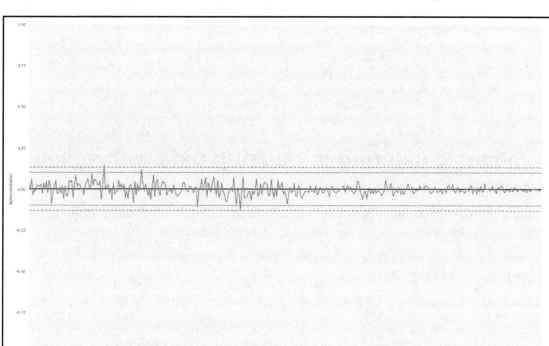

Monthly autocorrelation plot

There are no meaningful significant lags in this series, but if you try it with macroeconomic variables, such as the GDP, inflation rate, and unemployment level, you might observe significant quarterly or yearly peaks.

SciPy and scikit-learn

Scikit-learn is one of the SciKit libraries for machine learning, and it's built on top of SciPy. You can use it to perform regression analysis, as you've done in previous chapters with the scikit-learn library. Take a look at this code:

```
from sklearn import datasets, linear_model
from sklearn.metrics import mean_squared_error, r2_score

diabetes = datasets.load_diabetes()
```

```
linreg = linear_model.LinearRegression()

linreg.fit(diabetes.data, diabetes.target)

# You can inspect the results by looking at evaluation metrics
print('Coeff.: n', linreg.coef_)
print("MSE: {}".format(mean_squared_error(diabetes.target,
linreg.predict(diabetes.data)))) print('Variance Score:
{}'.format(r2_score(diabetes.target, linreg.predict(diabetes.data))))
```

K-means clustering in housing data with scikit-learn

In this section, we will cluster housing data with scikit-learn's k-means algorithm, as shown here:

```
from sklearn.cluster import KMeans from sklearn.datasets import load_boston
boston = load_boston()
# As previously, you have implemented the KMeans from scracth and in this
example, you use sklearns API k_means = KMeans(n_clusters=3) # Training
k_means.fit(boston.data)
KMeans(algorithm='auto', copy_x=True, init='k-means++', max_iter=300,
n_clusters=3, n_init=10, n_jobs=1, precompute_distances='auto',
random_state=None, tol=0.0001, verbose=0)
print(k_means.labels_)
```

The output of the preceding code is as follows:

```
[1 1 1 1 1 1 1 1 1 1 1 1 1 1 1 1 1 1 1 1 1 1 1 1 1 1 1 1 1 1 1 1 1 1 1 1 1 1 1 1 1
 1 1 1 1 1 1 1 1 1 1 1 1 1 1 1 1 1 1 1 1 1 1 1 1 1 1 1 1 1 1 1 1 1 1 1 1 1 1 1
 1 1 1 1 1 1 1 1 1 1 1 1 1 1 1 1 1 1 1 1 1 1 1 1 1 1 1 1 1 1 1 0 1 1 1 1 1 1 1
 1 1 1 1 1 1 1 1 1 1 1 1 1 1 1 1 1 1 1 1 1 1 1 1 1 1 1 1 1 1 1 1 1 1 1 1 1 1 1
 1 1 1 1 1 1 1 0 0 1 1 1 1 1 1 1 1 1 1 1 1 1 1 1 1 1 1 1 1 1 1 1 1 1 1 1 1 1 1
 1 1 1 1 1 1 1 1 1 1 1 1 1 1 1 1 1 1 1 1 1 1 1 1 1 1 1 1 1 1 1 1 1 1 1 1 1 1 1
 1 1 1 1 1 1 1 1 1 1 1 1 1 1 1 1 1 1 1 1 1 1 1 1 1 1 1 1 1 1 1 1 1 1 1 1 1 1 1
 1 1 1 1 1 1 1 1 1 1 1 1 1 1 1 1 1 1 1 1 1 1 1 1 1 1 1 1 1 1 1 1 1 1 1 1 1 1 1
 1 1 1 1 1 1 1 1 1 1 1 1 1 1 1 1 1 1 1 1 1 1 1 1 1 1 1 1 1 1 1 1 1 1 1 1 1 1 1
 1 1 1 1 1 1 1 1 1 1 1 1 1 1 1 1 1 1 1 1 1 2 2 2 2 2 2 2 2 2 2 0 2 2
 2 2 2 2 2 2 2 2 2 2 2 2 2 2 2 2 2 2 2 2 2 2 2 2 2 2 2 2 2 2 2 2 2 2
 2 2 0 0 0 0 0 0 0 0 0 2 2 0 0 0 0 0 0 0 0 0 0 0 0 0 0 0 0 2 2 2 2 2
 2 0 2 2 2 0 2 2 0 0 0 2 2 2 2 2 2 2 0 2 2 2 2 2 2 2 2 2 2 2
 2 2 2 2 2 2 2 2 2 2 2 1 1 1 1 1 1 1 1 1 1 1 1 1]
```

You can examine the cluster centers, using this line of code:

```
print(k_means.cluster_centers_)
```

This is the console's output:

```
[[ 1.49558803e+01 -5.32907052e-15 1.79268421e+01 2.63157895e-02
6.73710526e-01 6.06550000e+00 8.99052632e+01 1.99442895e+00
2.25000000e+01 6.44736842e+02 1.99289474e+01 5.77863158e+01
2.04486842e+01]
 [ 3.74992678e-01 1.57103825e+01 8.35953552e+00 7.10382514e-02
5.09862568e-01 6.39165301e+00 6.04133880e+01 4.46074481e+00
4.45081967e+00 3.11232240e+02 1.78177596e+01 3.83489809e+02
1.03886612e+01]
 [ 1.09105113e+01 5.32907052e-15 1.85725490e+01 7.84313725e-02
6.71225490e-01 5.98226471e+00 8.99137255e+01 2.07716373e+00
2.30196078e+01 6.68205882e+02 2.01950980e+01 3.71803039e+02
1.78740196e+01]]
```

In terms of the evaluation of clustering algorithms, you will usually use techniques such as silhouette analysis or the elbow method to assess the quality of clusters and to determine the right hyperparameters (such as the k for k-means). With the simple API that scikit-learn provides, you will also find such analysis easy to perform. You are strongly encouraged to build on what you have learned from these examples through practice, which will improve your knowledge and skills.

Summary

In this chapter, you practiced NumPy, SciPy, Pandas, and scikit-learn, using various examples, mainly for machine learning tasks. When you use Python data science libraries, there is usually more than one way of performing given task, and it usually helps to know more than one method.

You can either use alternatives for better implementations or for the sake of comparison. While trying different methods for a given task, you may either find different options that will allow you to further customize the implementation or simply observe some performance improvements.

The aim of this chapter was to show you these different options, and how flexible the Python language is because of its rich ecosystem of analytics libraries. In the next chapter, you will learn more about NumPy internals, such as how numpy manages data structures and memory, code profiling, and also tips for efficient programming.

7
Advanced Numpy

Many libraries have nice and easy to use APIs. All you need to do is to invoke the provided API function and library will handle the rest of it for you. What's happening under the hood is not your concern and you are only interested in output. This is fine for most cases, however it's important to understand at least the basic internal of a library you are using. Understanding internals will help you to grasp what's going on with your code and what are the red flags that you should avoid while developing your applications.

In this chapter, NumPy internals will be reviewed such as NumPy's type hierarchy and memory usage. At the end of this chapter, you will learn about the code profiling to inspect your programs line by line.

NumPy internals

As you have seen in previous chapters, NumPy arrays make numerical computations efficient and its API is intuitive and easy to use. NumPy array are also core to other scientific libraries as many of them are built on top of NumPy arrays.

In order to write better and more efficient code, you need to understand the internals of data handling. A NumPy array and its metadata live in a data buffer, which is a dedicated block of memory with certain data items.

How does NumPy manage memory?

Once you initialize a NumPy array, its metadata and data are stored at allocated memory locations in **Random Access Memory (RAM)**.

```
import numpy as np
array_x = np.array([100.12, 120.23, 130.91])
```

First, Python is a dynamically typed languages; there is no need for the explicit declaration of variables types such as int or double. Variable types are inferred and you'd expect that in this case the data type of array_x is np.float64:

```
print(array_x.dtype)
float64
```

The advantage of using the numpy library rather than Python is that numpy supports many different numerical data types such as bool_, int_, intc, intp, int8, int16, int32, int64, uint8, uint16, uint32, uint64, float_, float16, float32, float64, complex_, complex64, and complex128.

You can see these types by checking sctypes:

```
np.sctypes
{'complex': [numpy.complex64, numpy.complex128, numpy.complex256],
 'float': [numpy.float16, numpy.float32, numpy.float64, numpy.float128],
 'int': [numpy.int8, numpy.int16, numpy.int32, numpy.int64],
 'others': [bool, object, bytes, str, numpy.void],
 'uint': [numpy.uint8, numpy.uint16, numpy.uint32, numpy.uint64]}
```

Following diagram shows the data type tree:

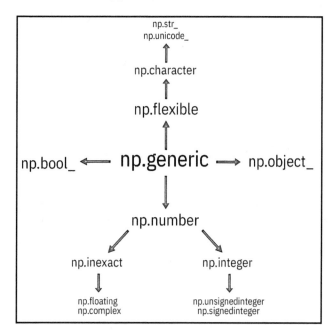

You can check the parent classes of data types such as `np.float64` by calling the `mro` method:

```
np.float64.mro()
[numpy.float64,
numpy.floating,
numpy.inexact,
numpy.number,
numpy.generic,
float,
object]
```

And parent classes for `np.int64`:

```
np.int64.mro()
[numpy.int64,
numpy.signedinteger,
numpy.integer,
numpy.number,
numpy.generic,
object]
```

`mro` method stands for Method Resolution Order. In order to understand it better, the concept of inheritance should be understood first. In programming languages where you can use the object oriented paradigm, you can base the properties and methods of an object upon another object that you have created before and that's what's called inheritance. In the previous example, `np.int64` retains the properties and behaviors of `np.signedinteger` and the ones that comes after that.

Let's see a simple example:

```
class First:
    def firstmethod(self):
        print("Call from First Class, first method.")

class Second:
    def secondmethod(self):
        print("Call from Second Class, second method.")

class Third(First, Second):
    def thirdmethod(self):
        print("Call from Third Class, third method.")
```

Here, there are 3 classes, while `First` and `Second` class are independent, `Third` class inherits from `First` and `Second`. You can create an instance of `Third` class and check it's contents with `dir` method:

```
myclass = Third()
dir(myclass)
[...
  '__repr__',
  '__setattr__',
  '__sizeof__',
  '__str__',
  '__subclasshook__',
  '__weakref__',
  'firstmethod',
  'secondmethod',
  'thirdmethod']
```

`dir` shows there are `firstmethod`, `secondmethod` and `thirdmethod` are among the methods of `myclass`.

You can call these methods as see their output as following:

```
myclass.firstmethod()
myclass.secondmethod()
myclass.thirdmethod()
# Call from First Class, first method.
# Call from Second Class, second method.
# Call from Third Class, third method.
```

Now, let's add a `firstmethod` to `Second` class and see what happens:

```
class First:
    def firstmethod(self):
        print("Call from First Class, first method.")

class Second:
    def firstmethod(self):
        print("Call from Second Class, first method.")
    def secondmethod(self):
        print("Call from Second Class, second method.")

class Third(First, Second):
    def thirdmethod(self):
        print("Call from Third Class, third method.")
```

Checking the method outputs as before:

```
myclass = Third()
myclass.firstmethod()
myclass.secondmethod()
myclass.thirdmethod()
# Call from First Class, first method.
# Call from Second Class, second method.
# Call from Third Class, third method.
```

As you can see, the method you have added to `Second` class has no effect, since an instance of `Third` class takes that from `First` class.

You can check the `mro` of classes as following:

```
Third.__mro__
```

Which will give you the following output:

```
(__main__.Third, __main__.First, __main__.Second, object)
```

This is how properties and methods are resolved when you use the inheritance mechanism and by now, you should know more or less how `mro` works. Now you can once again look at the `mro` examples for numpy data types we had before.

You can use nbytes to see the required memory to store data types.

First let's see the size of a single `float64`:

```
np.float64(100.12).nbytes
8
np.str_('n').nbytes
4
np.str_('numpy').nbytes
20
```

`array_x` has 3 `float64` and its size will be the number of elements multiplied by the item size, which is 24 as shown in the following snippet:

```
np.float64(array_x).nbytes
24
```

When you need less precision in your calculation you can for example use `np.float32`, which will take half of the memory that `float64` takes:

```
array_x2 = array_x.astype(np.float32)
array_x2
array([100.12, 120.23, 130.91], dtype=float32)
np.float32(array_x2).nbytes
12
```

Simply, 8 bytes of memory will hold 1 `float64` or 2 `float32`.

The dynamic nature of Python introduces a new way of handling data types as Python should contain more information about the data that it stores. While typical C variables will have the information about the memory the location, Python variables should have information stored as a C structure which contains reference count, the type of the object, object size, and the variable itself.

This is necessary to provide a flexible environment for working with different data types. If data structures such as list can hold different types of object this is due to the storage of this information for each element in the list.

However, since the data type in NumPy arrays is fixed, memory usage could be much more efficient due to using contiguous blocks of memory.

You can see the address and other related information by checking the __array_interface__ property of NumPy arrays.

This interface is written to allow developers to share the array memory and information:

```
array_x.__array_interface__
{'data': (140378873611440, False),
'descr': [('', '<f8')],
'shape': (3,),
'strides': None,
'typestr': '<f8',
'version': 3}
```

__array_interface__ is a python dictionary which has 6 keys:

- shape works like the usual shape attribute of NumPy array or pandas dataframe. It shows the size for each dimension. Since array_x has 1 dimension and 3 elements, it's tuple with size 3.
- typestr has 3 values, first showing the byte-order, the second showing the character-code and remaining characters showing the number of bytes. In this example, it has value '<f8' which means that byte-order is little-endian, character code is floating point and 8 is bytes used.
- descr may provide more detailed information about the memory layout. It's default value is [('', typestr)].
- data shows you where your data is stored. It's a tuple where first element shows the memory block address of NumPy array and second element is a flag which indicated whether it's read-only or not. In this example, memory block address is 140378873611440 and it's not read-only.
- strides indicates whether given array is C-style contiguous memory buffer. In this example None indicates that this is a C-style contiguous array. Otherwise, it will include tuple of strides to understand where to jump to retrieve next array element in a given dimension. Strides are important property as it guides array views when you use different slicing such as X[::4].
- version indicates version number which is 3 in this example.

Following snippet shows a quick example:

```
import numpy as np

X = np.array([1,2,3,2,1,3,9,8,11,12,10,11,14,25,26,24,30,22,24,27])

X[::4]
# array([ 1, 1, 11, 14, 30])
```

This is important to know because when you create new `ndarrays` by using slicing based on existing `ndarrays`, it may degrade the performance. Let's see a simple example; the following snippet creates a 3D `ndarray`:

```
nd_1 = np.random.randn(4, 6, 8)

nd_1
# array([[[ 0.64900179, -0.00494884, -0.97565618, -0.78851039],
[ 0.05165607, 0.068948 , 1.54542042, 1.68553396],
[-0.80311258, 0.95298682, -0.85879725, 0.67537715]],
[[ 0.24014811, -0.41894241, -0.00855571, 0.43483418],
[ 0.43001636, -0.75432657, 1.16955535, -0.42471807],
[ 0.6781286 , -1.87876591, 1.02969921, 0.43215107]]])
```

You can slice it and create another array:

```
nd_2 = nd_1[::, ::2, ::2]
```

Which will select:

1. First, all items from the first dimension
2. Then, every second item from the second dimension
3. Then, every second item from the third dimension

It will have the following array:

```
print(nd_2)
[[[ 0.64900179 -0.97565618]
[-0.80311258 -0.85879725]]
[[ 0.24014811 -0.00855571]
[ 0.6781286 1.02969921]]]
```

You can see that the memory address of `nd_1` and `nd_2` is the same:

```
nd_1.__array_interface__
{'data': (140547049888960, False),
'descr': [('', '<f8')],
'shape': (2, 3, 4),
```

```
'strides': None,
'typestr': '<f8',
'version': 3}

nd_2.__array_interface__
{'data': (140547049888960, False),
'descr': [('', '<f8')],
'shape': (2, 2, 2),
'strides': (96, 64, 16),
'typestr': '<f8',
'version': 3}
```

nd_2 has strides information to see how to move along a different dimension of the nd_1 array.

To emphasize the effect of these strides in numerical computations, the following example will use a larger size for array dimensions and slices.

```
nd_1 = np.random.randn(400, 600)
nd_2 = np.random.randn(400, 600*20)[::, ::20]
```

nd_1 and nd_2 have the same dimensions:

```
print(nd_1.shape, nd_2.shape)
(400, 600) (400, 600)
```

You can measure the time spent calculating the cumulative product of array elements for both nd_1 and nd_2:

```
%%timeit
np.cumprod(nd_1)
# 802 µs ± 20.2 µs per loop (mean ± std. dev. of 7 runs, 1000 loops each)

%%timeit
np.cumprod(nd_2)
# 12 ms ± 71.7 µs per loop (mean ± std. dev. of 7 runs, 100 loops each)
```

There's a significant time difference between two operations; why is that? As you might expect, strides in nd_2 cause this problem:

```
nd_1.__array_interface__
{'data': (4569473024, False),
'descr': [('', '<f8')],
'shape': (400, 600),
'strides': None,
'typestr': '<f8',
'version': 3}
```

```
nd_2.__array_interface__
{'data': (4603252736, False),
 'descr': [('', '<f8')],
 'shape': (400, 600),
 'strides': (96000, 160),
 'typestr': '<f8',
 'version': 3}
```

The presence of strides in nd_2 causes jumps to different memory locations when reading data from memory to CPU. If array elements are stored sequentially as a contiguous block of memory, then this operation is faster as seen from time measurements. Smaller strides are better to utilize CPU cache better for performance.

There are some workarounds to mitigate CPU cache related issues. One of the libraries that you can use is numexpr library (https://github.com/pydata/numexpr), which is a fast numerical expression evaluator for NumPy. Library makes memory usage more efficient and can also benefit multi-threaded programming to fully utilize available cores. You can also use it with Intel's VML for further improvements.

Let's see if it will help with nd_2 in following example:

```
import numexpr as ne

%%timeit
2 * nd_2 + 48
# 4 ms ± 10.9 µs per loop (mean ± std. dev. of 7 runs, 1000 loops each)

%%timeit
ne.evaluate("2 * nd_2 + 48")
# 843 µs ± 8.1 µs per loop (mean ± std. dev. of 7 runs, 1000 loops each)
```

You should try it with different examples to see the performance gains.

If you start indexing your array from the beginning up to some element, you will see that it has same memory address:

```
array_x[:2].__array_interface__['data'][0]
# 140378873611440
array_x[:2].__array_interface__['data'][0] ==
array_x.__array_interface__['data'][0]
# True
```

However, if you start indexing at a point other than 0, it will give you a different memory address:

```
array_x[1:].__array_interface__['data'][0]
# 140378873611448
array_x[1:].__array_interface__['data'][0] ==
array_x.__array_interface__['data'][0]
# False
```

There is another property of ndarray called flags which provides information about the memory layout of a given NumPy array:

```
array_f = np.array([[100.12, 120.23, 130.91], [90.45, 110.32, 120.32]])
print(array_f)
# [[100.12 120.23 130.91]
# [ 90.45 110.32 120.32]]

array_f.flags
# C_CONTIGUOUS : True
# F_CONTIGUOUS : False
# OWNDATA : True
# WRITEABLE : True
# ALIGNED : True
# WRITEBACKIFCOPY : False
# UPDATEIFCOPY : False
```

You can get individual flags either with dictionary-like notations or lower-case attribute names:

```
array_f.flags['C_CONTIGUOUS']
# True

array_f.flags.c_contiguous
# True
```

Let's have a look at each attribute:

- C_CONTIGUOUS: Single block of contiguous memory, C-style
- F_CONTIGUOUS: Single block of contiguous memory, Fortran-style

Your data can be stored using different layouts in memory. There are 2 different memory layouts to consider here: row major order which corresponds to C_CONTIGUOUS, and column major order, which corresponds to F_CONTIGUOUS.

In this example, `array_f` is 2-dimensional and the row items of `array_f` are stored in adjacent memory locations. Similarly, in `F_CONTIGUOUS` case, the values of each column are stored in adjacent memory locations.

Some `numpy` functions will take an argument `order` to indicate this order as `'C'` or `'F'`. Following example shows reshape function with different orders:

```
np.reshape(array_f, (3, 2), order='C')
# array([[100.12, 120.23],
# [130.91, 90.45],
# [110.32, 120.32]])

np.reshape(array_f, (3, 2), order='F')
# array([[100.12, 110.32],
# [ 90.45, 130.91],
# [120.23, 120.32]])
```

And the rest:

- `OWNDATA`: Whether the array shares its block of memory with another object or it has the ownership
- `WRITEABLE`: False means that it's read-only; otherwise this area could be written to.
- `ALIGNED`: Whether data's aligned for hardware
- `WRITEBACKIFCOPY`: Whether the array is a copy from another array or not
- `UPDATEIFCOPY`: (Deprecated, use `WRITEBACKIFCOPY`)

Understanding memory management matters since it affects the performance. Computation speed will be different depending on how you perform the calculations. You might not realize that some computations involve implicit copies of existing arrays, which would slow down computations.

Following code block shows 2 examples where the first doesn't require a copy whereas 2nd has an implicit copy operation:

```
shape = (400,400,400)

array_x = np.random.random_sample(shape)

import cProfile
import re

cProfile.run('array_x *= 2')
```

```
# 3 function calls in 0.065 seconds
# Ordered by: standard name
# ncalls tottime percall cumtime percall filename:lineno(function)
#      1   0.065   0.065   0.065   0.065 <string>:1(<module>)
#      1   0.000   0.000   0.065   0.065 {built-in method builtins.exec}
#      1   0.000   0.000   0.000   0.000 {method 'disable' of
'_lsprof.Profiler' objects}

import cProfile
import re
cProfile.run('array_y = array_x * 2')

# 3 function calls in 0.318 seconds
# Ordered by: standard name
# ncalls  tottime  percall  cumtime  percall filename:lineno(function)
#      1    0.318    0.318    0.318    0.318 <string>:1(<module>)
#      1    0.000    0.000    0.318    0.318 {built-in method
builtins.exec}
#      1    0.000    0.000    0.000    0.000 {method 'disable' of
'_lsprof.Profiler' objects}
```

First run is 5 times slower than the 2nd. You need to be aware of implicit copy operations and get familiar with in which situations it happens. Reshaping arrays does not require implicit copy unless it's a matrix transpose.

Many array operations return a new array for results. This behavior is expected but damages performance in iterative tasks where you could have millions or billions of iterations. Some numpy functions has out argument which creates output array and use it to write results of iterations. By this way, your program manage the memory better and requires less time:

```
shape_x = (8000,3000)

array_x = np.random.random_sample(shape_x)

%%timeit
np.cumprod(array_x)
# 176 ms ± 2.32 ms per loop (mean ± std. dev. of 7 runs, 10 loops each)
```

output_array should have the same type and dimensions as the expected output of the operations:

```
output_array = np.zeros(array_x.shape[0] * array_x.shape[1])

%%timeit
np.cumprod(array_x, out=output_array)
# 86.4 ms ± 1.21 ms per loop (mean ± std. dev. of 7 runs, 10 loops each)
```

Profiling NumPy code to understand the performance

There are couple of helpful libraries to monitor performance metrics of a given python script. You have already seen the usage of `cProfile` library. This section will introduce `vprof` which is visual profiler library.

It will provide you runtime statistics and memory utilization of a given python program.

1D clustering function from Chapter 5, *Clustering Clients of Wholesale Distributor Using NumPy,* will be used here and following code snippet should be saved to a file named `to_be_profiled.py`:

```python
import numpy as np

X = np.array([1,2,3,2,1,3,9,8,11,12,10,11,14,25,26,24,30,22,24,27])

n_clusters = 3

def Kmeans_1D(X, n_clusters, random_seed=442):

    # Randomly choose random indexes as cluster centers
    rng = np.random.RandomState(random_seed)
    i = rng.permutation(X.shape[0])[:n_clusters]
    c_centers = X[i]

    # Calculate distances between each point and cluster centers
    deltas = np.array([np.abs(point - c_centers) for point in X])

    # Get labels for each point
    labels = deltas.argmin(1)

    while True:

        # Calculate mean of each cluster
        new_c_centers = np.array([X[np.where(deltas.argmin(1) == i)[0]].mean()
for i in range(n_clusters)])

        # Calculate distances again
        deltas = np.array([np.abs(point - new_c_centers) for point in X])

        # Get new labels for each point
        labels = deltas.argmin(1)

        # If there's no change in centers, exit
```

```
    if np.all(c_centers == new_c_centers):
      break
    c_centers = new_c_centers

  return c_centers, labels

c_centers, labels = Kmeans_1D(X, 3)

print(c_centers, labels)
```

Once you save this file, you can start profiling it by using command line.

`vprof` can be configured in 4 different ways to get:

- CPU flame graph (`vprof -c c to_be_profiled.py`)
- Built-in profiler statistics (`vprof -c p to_be_profiled.py`)
- Memory graph for Cpython Garbage Collector tracked objects and process memory after lines in your programs are run (`vprof -c m to_be_profiled.py`)
- Code `heatmap` with `runtime` and count for each executed line (`vprof -c h to_be_profiled.py`)

These 4 configuration can be used either for single source file or a package. Let's see output for configurations p, m and h:

The configuration for the profiler:

```
$ vprof -c p to_be_profiled.py
Running Profiler...
[10.71428571 25.42857143 2. ] [2 2 2 2 2 0 0 0 0 0 0 0 1 1 1 1 1 1 1]
Starting HTTP server...
```

A new tab will be opened in your browser with the following output:

Color	%	Function name	Filename	Line	Time
	100.0253%	<built-in method builtins.exec>	~	0	0.1067s
	100.0169%	<module>	to_be_profiled.py	1	0.1067s
	98.9477%	_find_and_load		958	0.1056s
	98.8418%	_find_and_load_unlocked		931	0.1055s
	98.302%	_load_unlocked		641	0.1049s
	98.2008%	exec_module		672	0.1048s
	97.9347%	_call_with_frames_removed		197	0.1045s
	97.9319%	<module>	/Users/umitcakmak/anaconda/lib/python3.6/site-packages/numpy/__init...	106	0.1045s
	95.6726%	_handle_fromlist		989	0.1021s
	95.3999%	<built-in method builtins.__import__>		0	0.1018s
	77.29%	<module>	/Users/umitcakmak/anaconda/lib/python3.6/site-packages/numpy/add_ne...	10	0.0825s
	75.3587%	<module>	/Users/umitcakmak/anaconda/lib/python3.6/site-packages/numpy/lib/__...	1	0.0804s
	58.9364%	<module>	/Users/umitcakmak/anaconda/lib/python3.6/site-packages/numpy/lib/ty...	3	0.0629s
	58.3695%	<module>	/Users/umitcakmak/anaconda/lib/python3.6/site-packages/numpy/core/_...	1	0.0623s
	34.6287%	module_from_spec		553	0.037s
	31.9983%	create_module		919	0.0341s
	31.9486%	<built-in method _imp.create_dynamic>	~	0	0.0341s
	21.4946%	get_code		743	0.0229s
	14.6109%	<module>	/Users/umitcakmak/anaconda/lib/python3.6/site-packages/numpy/testin...	7	0.0156s
	13.7806%	<module>	/Users/umitcakmak/anaconda/lib/python3.6/site-packages/numpy/core/_...	6	0.0147s
	12.8426%	_compile_bytecode		485	0.0137s
	12.3619%	<built-in method marshal.loads>	~	0	0.0132s
	12.2363%	_find_spec		861	0.0131s
	10.931%	find_spec		1149	0.0117s
	10.8213%	_get_spec		1117	0.0115s
	10.3744%	<module>	/Users/umitcakmak/anaconda/lib/python3.6/unittest/__init__.py	45	0.0111s
	9.0793%	find_spec		1233	0.0097s
	8.1647%	<built-in method builtins.__build_class__>		0	0.0087s
	6.6917%	<module>	/Users/umitcakmak/anaconda/lib/python3.6/site-packages/numpy/compat...	10	0.0071s
	6.0929%	<module>	/Users/umitcakmak/anaconda/lib/python3.6/site-packages/numpy/compat...	4	0.0065s
	5.9392%	<module>	/Users/umitcakmak/anaconda/lib/python3.6/site-packages/numpy/random...	88	0.0063s
	5.2532%	<module>	/Users/umitcakmak/anaconda/lib/python3.6/site-packages/pathlib.py	1	0.0056s
	5.0312%	<module>	/Users/umitcakmak/anaconda/lib/python3.6/unittest/case.py	1	0.0054s
	4.6779%	<module>	/Users/umitcakmak/anaconda/lib/python3.6/site-packages/numpy/ma/__i...	41	0.005s
	4.5532%	<module>	/Users/umitcakmak/anaconda/lib/python3.6/site-packages/numpy/lib/po...	4	0.0049s
	4.4286%	get_data		830	0.0047s
	4.2037%	_compile	/Users/umitcakmak/anaconda/lib/python3.6/re.py	286	0.0045s
	4.1625%	_path_stat		75	0.0044s
	4.1035%	compile	/Users/umitcakmak/anaconda/lib/python3.6/re.py	231	0.0044s
	3.9338%	<built-in method posix.stat>	~	0	0.0042s
	3.9198%	<module>	/Users/umitcakmak/anaconda/lib/python3.6/site-packages/numpy/linalg...	45	0.0042s
	3.8935%	compile	/Users/umitcakmak/anaconda/lib/python3.6/sre_compile.py	557	0.0042s
	3.5871%	<module>	/Users/umitcakmak/anaconda/lib/python3.6/site-...	15	0.0038s

Object name: to_be_profiled.py (module)
Total time: 0.10671500000000013s
Primitive calls: 37162
Total calls: 38016

Time spent for each call

You can see the filename, function name, line number and time spent for each call.

Configuration for memory usage statistics:

```
$ vprof -c m to_be_profiled.py
Running MemoryProfiler...
[10.71428571 25.42857143 2. ] [2 2 2 2 2 2 0 0 0 0 0 0 0 1 1 1 1 1 1 1]
Starting HTTP server...
```

New tab will be opened in your browser with following output:

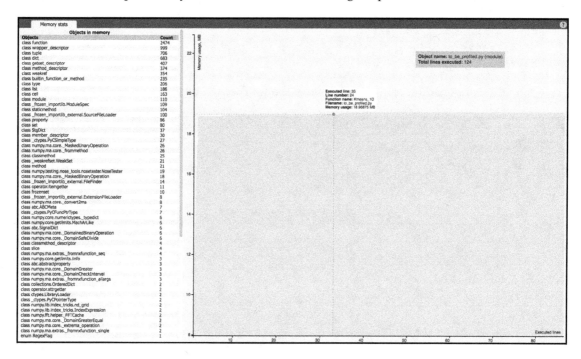

Memory usage

On the left, you can see the objects in memory and the chart shows you the memory usage in megabytes as the number of executed lines increases. If you hover your mouse over the chart, every point will have the following information:

- Executed line
- Line number
- Function name
- Filename
- Memory usage

For example line 27 in `to_be_profiled.py` file is following line which calculates `deltas`:

```
deltas = np.array([np.abs(point - new_c_centers) for point in X])
```

It's executed many time since it's a list comprehension if you inspect the chart.

Configuration for code `heatmap`:

```
$ vprof -c h to_be_profiled.py
Running CodeHeatmapProfiler...
[10.71428571 25.42857143 2. ] [2 2 2 2 2 2 0 0 0 0 0 0 0 1 1 1 1 1 1 1]
Starting HTTP server...
```

New tab will be opened in your browser with the following output:

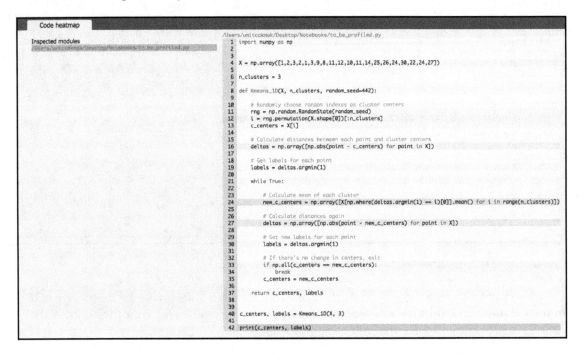

Heatmap for the lines executed

On the left, there is a list of inspected module and there is only one file to be inspected in this case.

On the right, there's a heatmap of every line included in your program. If you hover your mouse over lines, it will give you the following information:

- **Time spent**
- **Total running time**
- **Percentage**
- **Run count**

If you hover your mouse over line `27`, it will give you the following statistics:

```
26        # Calculate distances again
27        deltas = np.array([np.abs(point - new_c_centers) for point in X])
28     Time spent: 0.0002586841583251953 s
29     Total running time: 0.0027663707733154297 s
30     Percentage: 9.35%
31     Run count: 66
```

Summary

Knowing NumPy's internals is crucially important when you are working with scientific operations. Efficiency is key since many scientific computations are compute and memory intensive. Hence, if your code is not written efficiently, computations will take much longer than they need and this will hurt your research and development timeline.

In this chapter, you have seen some of the internals and performance aspects of the NumPy library and also learned about the `vprof` library, which helps you inspect the performance of your python programs.

Code profiling will help you a lot to inspect your programs line by line and there are different ways of looking at the same data, as you have seen previously. Once you have identified the most demanding parts of your programs, then you can start searching for more efficient ways or implementations to improve performance and save more time.

In the next chapter, we'll take an overview of high-performance, low-level numerical computing libraries. NumPy can use these implementations for considerable performance gains.

8
Overview of High-Performance Numerical Computing Libraries

There are many numerical operations that can be performed in scientific computing applications, and non-optimized code or library implementations cause serious performance bottlenecks.

The NumPy library helps to increase the performance of Python programs by using its memory layout more efficiently.

One of the most commonly used branches of mathematics in real-world applications is linear algebra. Linear algebra is used for computer graphics, cryptography, econometrics, machine learning, deep learning, and natural language processing, to name but a few of its uses. Having performant matrix and vector operations is crucial.

High-performance, low-level frameworks, such as BLAS, LAPACK, and ATLAS—which are part of Netlib's libraries, and are used for dense linear algebra operations—and other frameworks, such as Intel MKL, are there for you to use in your programs. These libraries are highly performant and accurate in their calculations. You can use these libraries by calling them with other high-level programming languages, such as Python or C++.

When NumPy is linked against different BLAS libraries, you can observe performance differences without changing your code and it's important to understand which linkage will better improve the performance.

Let's have a look at these low-level libraries.

BLAS and LAPACK

BLAS stands for **Basic Linear Algebra Subprograms**, and is a standard for dealing with low-level routines for linear algebra operations. Low-level routines include operations such as vector and matrix addition/multiplication, linear combinations, and so on. BLAS provides three different levels for linear algebra operations:

- **BLAS1**: Scalar–vector and vector–vector operations
- **BLAS2**: Matrix–vector operations
- **BLAS3**: Matrix–matrix operations

LAPACK stands for **Linear Algebra Package**, and contains higher-level operations. LAPACK provides routines for matrix factorizations—such as LU, Cholesky, and QR—and for solving eigenvalue problems. LAPACK mostly depends on BLAS routines.

ATLAS

There are many optimized BLAS implementations. **ATLAS** stands for **Automatically Tuned Linear Algebra Software**, and is a platform-independent project that generates an optimized BLAS implementation.

Intel Math Kernel Library

Intel MKL optimizes BLAS for Intel processors. There are improved routines and functions such as Level 1, 2 and 3 BLAS, LAPACK routines, solvers, FFT functions, other math and statistical functions. These improved routines and functions benefit from improvements like shared memory multiprocessing and they are used to accelerate scientific python libraries such as NumPy and SciPy in distributions such as Anaconda Distribution. If you look at its release notes (`https://software.intel.com/en-us/articles/intel-math-kernel-library-release-notes-and-new-features`), you can see several important improvements were made with every release, such as the improved performance of LAPACK functions.

OpenBLAS

OpenBLAS is another optimized BLAS library, and it provides BLAS3-level optimizations for different configurations. Authors reported performance enhancements and improvements over BLAS that were comparable with Intel MKL's quality of performance.

Configuring NumPy with low-level libraries using AWS EC2

1. Log in to AWS. If you don't have an account, create one:

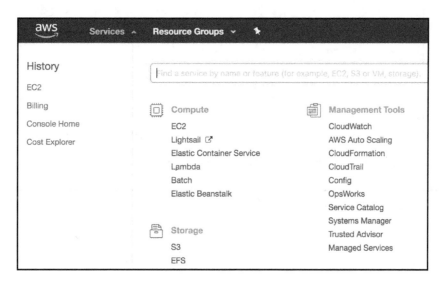

2. Select **EC2**.

3. Click **Launch Instance**:

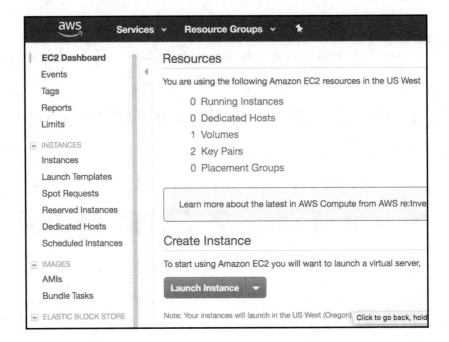

4. Select **Ubuntu Server 16.04 LTS (HVM), SSD Volume Type - ami-db710fa3**:

5. Select the **t2.micro** instance type:

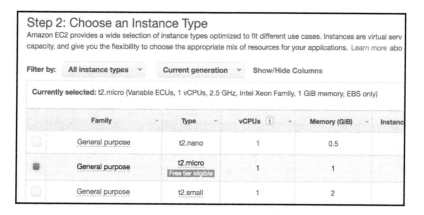

6. Click **Review and Launch**:

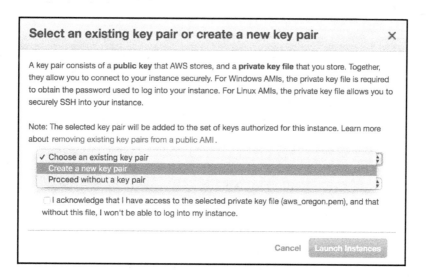

7. Click **Launch**.
8. Select **Create a new key pair**:

9. Give it a name and click **Launch Instances**. It will take a while for it to run:

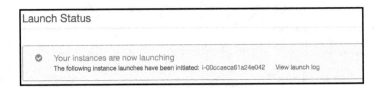

10. Once its status is `running`, click the **Instance ID**, which in this case is `i-00ccaeca61a24e042`. Then select the instance and click `Connect`:

11. It will then show you the following window with some useful information:

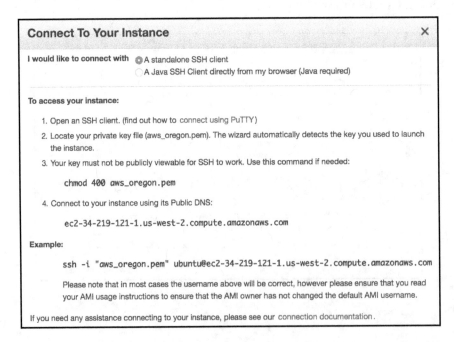

12. Open your terminal and navigate to the folder where you saved your generated key. The key name in this example is `aws_oregon`. Run the following command:

    ```
    $ chmod 400 aws_oregon.pem
    ```

13. Then, copy the line in the example section of the previous window and run it:

    ```
    $ ssh -i "aws_oregon.pem" ubuntu@ec2-34-219-121-1.us-
    west-2.compute.amazonaws.com
    ```

14. Type `yes` in answer to the first question to add it to the known hosts, and it will connect to your instance:

```
The authenticity of host 'ec2-34-219-121-1.us-west-2.compute.amazonaws.com (34.219.121.1)' can't be established.
ECDSA key fingerprint is SHA256:Mhxlf76E7CmSlNH52X4ls2EKeujAYYh4NETAfju9+cA.
Are you sure you want to continue connecting (yes/no)? yes
Warning: Permanently added 'ec2-34-219-121-1.us-west-2.compute.amazonaws.com,34.219.121.1' (ECDSA) to the list of known hosts.
Welcome to Ubuntu 16.04.4 LTS (GNU/Linux 4.4.0-1060-aws x86_64)

 * Documentation:  https://help.ubuntu.com
 * Management:     https://landscape.canonical.com
 * Support:        https://ubuntu.com/advantage

  Get cloud support with Ubuntu Advantage Cloud Guest:
    http://www.ubuntu.com/business/services/cloud

0 packages can be updated.
0 updates are security updates.



The programs included with the Ubuntu system are free software;
the exact distribution terms for each program are described in the
individual files in /usr/share/doc/*/copyright.

Ubuntu comes with ABSOLUTELY NO WARRANTY, to the extent permitted by
applicable law.

To run a command as administrator (user "root"), use "sudo <command>".
See "man sudo_root" for details.

ubuntu@ip-172-31-21-32:~$
```

The first thing that you need to do is update and upgrade the preinstalled packages by running the following commands:

```
sudo apt-get update
sudo apt-get upgrade
```

The `sudo` phrase gives you the necessary rights to update and upgrade, since changes in software packages may negatively affect the system, and not everybody should be able to authorize them. You can think of `apt-get` as the package manager for Ubuntu.

You can create many virtual environments with links to different low-level libraries, however here you will start with a new provisioned instance every time you are configuring NumPy with new low-level library. This will give you an idea about the configuration process which will later allow you to set-up your virtual environments with ease.

Installing BLAS and LAPACK

In order to set up your development environment, you need to install the required packages, such as the compilers, libraries, and other necessary pieces after running following commands,

```
$ sudo apt-get update
$ sudo apt-get upgrade
```

For this configuration, you are lucky as you can run the following command to install Python's SciPy package, and it will install all the required packages, including NumPy, Basic Linear Algebra Subprograms (`libblas3`), and Linear Algebra Package (`liblapack3`):

```
$ sudo apt-get install python3-scipy
```

Console output:

```
ubuntu@ip-172-31-21-32:/$ sudo apt-get install python3-scipy
Reading package lists... Done
Building dependency tree
Reading state information... Done
The following additional packages will be installed:
  binutils cpp cpp-5 g++ g++-5 gcc gcc-5 libasan2 libatomic1 libblas-common libblas3 libc-dev-bin libc6-dev libcc1-0 libcilkrts5
  libgcc-5-dev libgfortran3 libgomp1 libisl15 libitm1 liblapack3 liblsan0 libmpc3 libmpx0 libquadmath0 libstdc++-5-dev libtsan0
  libubsan0 linux-libc-dev manpages-dev python3-decorator python3-numpy
Suggested packages:
  binutils-doc cpp-doc gcc-5-locales g++-multilib g++-5-multilib gcc-5-doc libstdc++6-5-dbg gcc-multilib make autoconf automake
  libtool flex bison gdb gcc-doc gcc-5-multilib libgcc1-dbg libgomp1-dbg libitm1-dbg libatomic1-dbg libasan2-dbg liblsan0-dbg
  libtsan0-dbg libubsan0-dbg libcilkrts5-dbg libmpx0-dbg libquadmath0-dbg glibc-doc libstdc++-5-doc gfortran python-numpy-doc
  python3-dev python3-nose python3-numpy-dbg python-scipy-doc
The following NEW packages will be installed:
  binutils cpp cpp-5 g++ g++-5 gcc gcc-5 libasan2 libatomic1 libblas-common libblas3 libc-dev-bin libc6-dev libcc1-0 libcilkrts5
  libgcc-5-dev libgfortran3 libgomp1 libisl15 libitm1 liblapack3 liblsan0 libmpc3 libmpx0 libquadmath0 libstdc++-5-dev libtsan0
  libubsan0 linux-libc-dev manpages-dev python3-decorator python3-numpy python3-scipy
0 upgraded, 33 newly installed, 0 to remove and 0 not upgraded.
Need to get 49.8 MB of archives.
After this operation, 190 MB of additional disk space will be used.
Do you want to continue? [Y/n]
```

1. Type Y and press *Enter* to continue. Once the installation is complete, run the following command to open the `python3` interpreter:

```
$ python3
```

Starting Python console:

```
ubuntu@ip-172-31-21-32:~$ python3
Python 3.5.2 (default, Nov 23 2017, 16:37:01)
[GCC 5.4.0 20160609] on linux
Type "help", "copyright", "credits" or "license" for more information.
>>>
```

2. Import `numpy` and use the `show_config` method to see NumPy's configuration:

Console output:

```
>>> import numpy as np
>>> np.show_config()
openblas_lapack_info:
  NOT AVAILABLE
atlas_3_10_blas_threads_info:
  NOT AVAILABLE
atlas_blas_info:
  NOT AVAILABLE
atlas_threads_info:
  NOT AVAILABLE
atlas_3_10_info:
  NOT AVAILABLE
lapack_info:
    language = f77
    libraries = ['lapack', 'lapack']
    library_dirs = ['/usr/lib']
atlas_blas_threads_info:
  NOT AVAILABLE
blas_info:
    language = c
    libraries = ['blas', 'blas']
    library_dirs = ['/usr/lib']
    define_macros = [('HAVE_CBLAS', None)]
lapack_opt_info:
    define_macros = [('NO_ATLAS_INFO', 1), ('HAVE_CBLAS', None)]
    libraries = ['lapack', 'lapack', 'blas', 'blas']
    library_dirs = ['/usr/lib']
    language = c
atlas_info:
  NOT AVAILABLE
openblas_info:
  NOT AVAILABLE
blas_opt_info:
    define_macros = [('NO_ATLAS_INFO', 1), ('HAVE_CBLAS', None)]
    libraries = ['blas', 'blas']
    library_dirs = ['/usr/lib']
    language = c
lapack_mkl_info:
  NOT AVAILABLE
atlas_3_10_blas_info:
  NOT AVAILABLE
mkl_info:
  NOT AVAILABLE
blas_mkl_info:
  NOT AVAILABLE
atlas_3_10_threads_info:
  NOT AVAILABLE
```

3. As the BLAS and LAPACK libraries were available when NumPy was installing, it used them to build the library. You can see them in `lapack_info` and `blas_info`; otherwise, you won't see them in the output, as shown in the following screenshot:

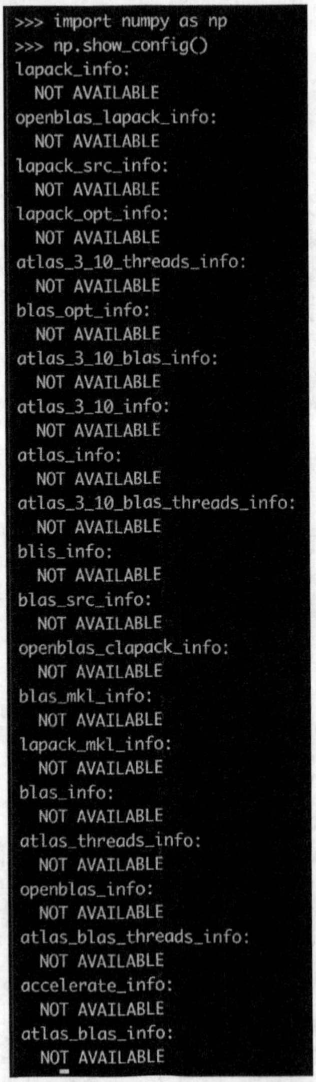

```
>>> import numpy as np
>>> np.show_config()
lapack_info:
  NOT AVAILABLE
openblas_lapack_info:
  NOT AVAILABLE
lapack_src_info:
  NOT AVAILABLE
lapack_opt_info:
  NOT AVAILABLE
atlas_3_10_threads_info:
  NOT AVAILABLE
blas_opt_info:
  NOT AVAILABLE
atlas_3_10_blas_info:
  NOT AVAILABLE
atlas_3_10_info:
  NOT AVAILABLE
atlas_info:
  NOT AVAILABLE
atlas_3_10_blas_threads_info:
  NOT AVAILABLE
blis_info:
  NOT AVAILABLE
blas_src_info:
  NOT AVAILABLE
openblas_clapack_info:
  NOT AVAILABLE
blas_mkl_info:
  NOT AVAILABLE
lapack_mkl_info:
  NOT AVAILABLE
blas_info:
  NOT AVAILABLE
atlas_threads_info:
  NOT AVAILABLE
openblas_info:
  NOT AVAILABLE
atlas_blas_threads_info:
  NOT AVAILABLE
accelerate_info:
  NOT AVAILABLE
atlas_blas_info:
  NOT AVAILABLE
```

4. If you are working in macOS, you will be able to utilize the Accelerate/vecLib framework. The following command will output the accelerator options of the macOS system:

```
>>> import numpy as np
>>> np.show_config()
blas_mkl_info:
  NOT AVAILABLE
blis_info:
  NOT AVAILABLE
openblas_info:
  NOT AVAILABLE
atlas_3_10_blas_threads_info:
  NOT AVAILABLE
atlas_3_10_blas_info:
  NOT AVAILABLE
atlas_blas_threads_info:
  NOT AVAILABLE
atlas_blas_info:
  NOT AVAILABLE
blas_opt_info:
    extra_compile_args = ['-msse3', '-I/System/Library/Frameworks/vecLib.framework/Headers']
    extra_link_args = ['-Wl,-framework', '-Wl,Accelerate']
    define_macros = [('NO_ATLAS_INFO', 3), ('HAVE_CBLAS', None)]
lapack_mkl_info:
  NOT AVAILABLE
openblas_lapack_info:
  NOT AVAILABLE
openblas_clapack_info:
  NOT AVAILABLE
atlas_3_10_threads_info:
  NOT AVAILABLE
atlas_3_10_info:
  NOT AVAILABLE
atlas_threads_info:
  NOT AVAILABLE
atlas_info:
  NOT AVAILABLE
lapack_opt_info:
    extra_compile_args = ['-msse3']
    extra_link_args = ['-Wl,-framework', '-Wl,Accelerate']
    define_macros = [('NO_ATLAS_INFO', 3), ('HAVE_CBLAS', None)]
```

Installing OpenBLAS

OpenBLAS has a slightly different steps as shown as follows:

1. Run the commands as follows in previous configurations:

```
$ sudo apt-get update
$ sudo apt-get upgrade
```

2. You need to install the build-essential which includes make command and other necessary libraries by running following command:

```
$ sudo apt-get install build-essential libc6 gcc gfortran
```

3. Create a file called `openblas_setup.sh` and paste the following content (`https://github.com/shivaram/matrix-bench/blob/master/build-openblas-ec2-usr-lib.sh`). You can find different setup scripts if you search GitHub and you can try the one which suits to your needs:

```
#!/bin/bash

set -e

pushd /root
git clone https://github.com/xianyi/OpenBLAS.git

pushd /root/OpenBLAS
  make clean
  make -j4

  rm -rf /root/openblas-install
  make install PREFIX=/root/openblas-install
popd

ln -sf /root/openblas-install/lib/libopenblas.so
/usr/lib/libblas.so
ln -sf /root/openblas-install/lib/libopenblas.so
/usr/lib/libblas.so.3
ln -sf /root/openblas-install/lib/libopenblas.so
/usr/lib/liblapack.so.3
```

4. Save this file and run the following commands:

```
$ chmod +777 openblas_setup.sh
$ sudo ./openblas_setup.sh
```

5. Once installation is done, you can install numpy and scipy as following:

```
$ sudo apt-get install python3-pip
$ pip3 install numpy
$ pip3 install scipy
```

6. Now, you can check the NumPy config as you did previously:

```
Python 3.5.2 (default, Nov 23 2017, 16:37:01)
[GCC 5.4.0 20160609] on linux
Type "help", "copyright", "credits" or "license" for more information.
>>> import numpy as np
>>> np.show_config()
blis_info:
  NOT AVAILABLE
lapack_mkl_info:
  NOT AVAILABLE
blas_mkl_info:
  NOT AVAILABLE
blas_opt_info:
    language = c
    libraries = ['openblas', 'openblas']
    define_macros = [('HAVE_CBLAS', None)]
    library_dirs = ['/usr/local/lib']
openblas_info:
    language = c
    libraries = ['openblas', 'openblas']
    define_macros = [('HAVE_CBLAS', None)]
    library_dirs = ['/usr/local/lib']
openblas_lapack_info:
    language = c
    libraries = ['openblas', 'openblas']
    define_macros = [('HAVE_CBLAS', None)]
    library_dirs = ['/usr/local/lib']
lapack_opt_info:
    language = c
    libraries = ['openblas', 'openblas']
    define_macros = [('HAVE_CBLAS', None)]
    library_dirs = ['/usr/local/lib']
```

Installing Intel MKL

In order to build NumPy and SciPy against Intel MKL, please follow these instructions

1. Run the commands as follows:

```
$ sudo apt-get update
$ sudo apt-get upgrade
```

2. You need to install Anaconda distribution since Anaconda installation comes with Intel MKL. First download the Anaconda with following command:

```
$ wget
https://repo.continuum.io/archive/Anaconda3-5.2.0-Linux-x86_64.sh
```

3. When installation is done, cd into anaconda3/bin and run python:

```
$ cd anaconda3/bin
$ ./python
```

4. You can check the numpy config as previously:

```
ubuntu@ip-172-31-22-134:~/anaconda3/bin$ ./python
Python 3.6.5 |Anaconda, Inc.| (default, Apr 29 2018, 16:14:56)
[GCC 7.2.0] on linux
Type "help", "copyright", "credits" or "license" for more information.
>>> import numpy as np
>>> np.show_config()
mkl_info:
    libraries = ['mkl_rt', 'pthread']
    library_dirs = ['/home/ubuntu/anaconda3/lib']
    define_macros = [('SCIPY_MKL_H', None), ('HAVE_CBLAS', None)]
    include_dirs = ['/home/ubuntu/anaconda3/include']
blas_mkl_info:
    libraries = ['mkl_rt', 'pthread']
    library_dirs = ['/home/ubuntu/anaconda3/lib']
    define_macros = [('SCIPY_MKL_H', None), ('HAVE_CBLAS', None)]
    include_dirs = ['/home/ubuntu/anaconda3/include']
blas_opt_info:
    libraries = ['mkl_rt', 'pthread']
    library_dirs = ['/home/ubuntu/anaconda3/lib']
    define_macros = [('SCIPY_MKL_H', None), ('HAVE_CBLAS', None)]
    include_dirs = ['/home/ubuntu/anaconda3/include']
lapack_mkl_info:
    libraries = ['mkl_rt', 'pthread']
    library_dirs = ['/home/ubuntu/anaconda3/lib']
    define_macros = [('SCIPY_MKL_H', None), ('HAVE_CBLAS', None)]
    include_dirs = ['/home/ubuntu/anaconda3/include']
lapack_opt_info:
    libraries = ['mkl_rt', 'pthread']
    library_dirs = ['/home/ubuntu/anaconda3/lib']
    define_macros = [('SCIPY_MKL_H', None), ('HAVE_CBLAS', None)]
    include_dirs = ['/home/ubuntu/anaconda3/include']
```

Installing ATLAS

In order to build NumPy against ATLAS, please follow these instructions

1. Run the commands as follows:

   ```
   $ sudo apt-get update
   $ sudo apt-get upgrade
   ```

2. You need to install the `build-essential` which includes make command and other necessary libraries by running following command:

   ```
   $ sudo apt-get install build-essential libc6 gcc gfortran
   ```

3. You then need to install `atlas`:

   ```
   $ sudo apt-get install libatlas-base-dev
   ```

4. You can now install `pip` and `numpy` as follows:

   ```
   $ sudo apt-get install python3-pip
   $ pip3 install --no-cache-dir Cython
   $ git clone https://github.com/numpy/numpy.git
   $ cd numpy
   $ cp site.cfg.example site.cfg
   $ vi site.cfg
   ```

   Inside `site.cfg`, you should comment out atlas lines and set it to your atlas installation as following:

   ```
   [atlas]
   library_dirs = /usr/local/atlas/lib
   include_dirs = /usr/local/atlas/include
   ```

   Then run:

   ```
   $ sudo python3 setup.py install
   ```

5. Once installation is complete, install `scipy`:

   ```
   $ pip3 install scipy
   ```

Then return to you home directory, start `python` interpreter and check `numpy` config, which will give you the following output:

```
>>> import numpy as np
>>> np.show_config()
atlas_blas_info:
 include_dirs = ['/usr/include/atlas']
 language = c
 library_dirs = ['/usr/lib/atlas-base']
 define_macros = [('HAVE_CBLAS', None), ('ATLAS_INFO', '"\\"3.10.2\\""')]
 libraries = ['f77blas', 'cblas', 'atlas', 'f77blas', 'cblas']
 ...
```

You have covered the configuration for all the low-level libraries mentioned. Time to understand the compute-intensive tasks for the benchmark.

Compute-intensive tasks for benchmarking

Now, you'll be able to benchmark NumPy performance using different configurations, such as with or without BLAS/LAPACK, OpenBLAS, ATLAS, and Intel MKL. Let's review what you are going to compute for benchmarks.

Matrix decomposition

Matrix decomposition, or **factorization methods**, involves calculating the constituents of a matrix so that they can be used to simplify more demanding matrix operations. In practice, this means breaking the matrix you have into more than one matrix so that, when you calculate the product of these smaller matrices, you get your original matrix back. Some examples of matrix decomposition methods are **singular-value decomposition** (**SVD**), eigenvalue decomposition, Cholesky decomposition, **lower–upper** (**LU**), and QR decomposition.

Singular-value decomposition

SVD is one of the most useful tools in linear algebra. Beltrami and Jordan published several papers on its use. SVD is used in a wide variety of applications, such as computer vision and signal processing.

If you have a square or rectangular matrix (*M*), you can decompose it into matrix (*U*), matrix (*V*) (using the transpose of the matrix in the calculation), and the singular value (*d*).

Your ultimate formula will look like the following:

$$M = U_1 d_1 V_1^T + \ldots + U_n d_n V_n^T$$

The following is an illustration of singular-value decomposition:

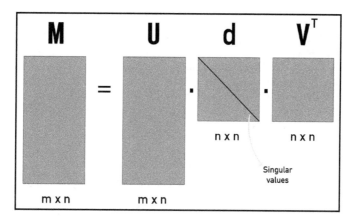

A simple data-reduction method would be to exclude parts of this formula where *d* is small enough to make that part negligible.

Let's see how this is implemented using numpy:

```
import numpy as np
M = np.random.randint(low=0, high=20, size=20).reshape(4,5)
print(M)

# Output
[[18 15 11 13 19]
 [ 1  6  8 13 18]
 [ 9  7 15 13 10]
 [17 15 12 14 12]]

U, d, VT = np.linalg.svd(M)
print("U:n {}n".format(U))
print("d:n {}n".format(d))
print("VT:n {}".format(VT))

U:
 [[-0.60773852 -0.22318957  0.5276743  -0.54990921]
```

```
  [-0.38123886  0.86738201  0.19333528  0.25480749]
  [-0.42657252  0.10181457 -0.82343563 -0.36003255]
  [-0.55076919 -0.43297652 -0.07832665  0.70925987]]
d:
 [56.31276456 13.15721839  8.08763849  2.51997135]
VT:
 [[-0.43547429 -0.40223663 -0.40386674 -0.46371223 -0.52002929]
[-0.72920427 -0.29835313  0.06197899  0.27638212  0.54682545]
 [ 0.11733943  0.26412864 -0.73449806 -0.30022507  0.53557916]
 [-0.32795351  0.55511623 -0.3571117   0.56067806 -0.3773643 ]
 [-0.39661218  0.60932187  0.40747282 -0.55144258  0.03609177]]


# Setting full_matrices to false gives you reduced form where small values
close to zero are excluded
U, d, VT = np.linalg.svd(M, full_matrices=False)
print("U:n {}n".format(U))
print("d:n {}n".format(d))
print("VT:n {}".format(VT))


# Output
U:
 [[-0.60773852 -0.22318957  0.5276743  -0.54990921]
  [-0.38123886  0.86738201  0.19333528  0.25480749]
  [-0.42657252  0.10181457 -0.82343563 -0.36003255]
  [-0.55076919 -0.43297652 -0.07832665  0.70925987]]
d:
 [56.31276456 13.15721839  8.08763849  2.51997135]
VT:
 [[-0.43547429 -0.40223663 -0.40386674 -0.46371223 -0.52002929]
  [-0.72920427 -0.29835313  0.06197899  0.27638212  0.54682545]
  [ 0.11733943  0.26412864 -0.73449806 -0.30022507  0.53557916]
  [-0.32795351  0.55511623 -0.3571117   0.56067806 -0.3773643 ]]
```

Cholesky decomposition

If you have a square matrix, you can also apply Cholesky decomposition, where you decompose a matrix (M) into two triangular matrices (U and U^T). Cholesky decomposition helps you to simplify computational complexity. It can be summed up in the following formula:

$$M = U^T U$$

The following is an illustration of Cholesky decomposition:

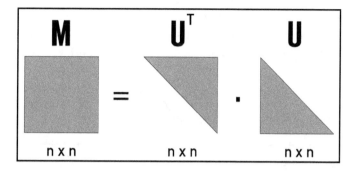

Let's see how it's implemented using numpy:

```
from numpy import array
from scipy.linalg import cholesky
M = np.array([[1, 3, 4],
[2, 13, 15],
[5, 31, 33]])

print(M)
# Output
[[ 1  3  4]
 [ 2 13 15]
 [ 5 31 33]]

L = cholesky(M)
print(L)

# Output
[[1.          3.          4.         ]
 [0.          2.          1.5        ]
 [0.          0.          3.84057287]]

L.T.dot(L)
# Output
array([[ 1.,  3.,  4.],
       [ 3., 13., 15.],
       [ 4., 15., 33.]])
```

Lower-upper decomposition

Similar to Cholesky decomposition, LU decomposition decomposes a matrix (*M*) into **lower** (*L*) and **upper** (*U*) triangular matrices. This also helps us to simplify computationally intensive algebra. It can be summed up in the following formula:

$$M=LU$$

The following is an illustration of LU decomposition:

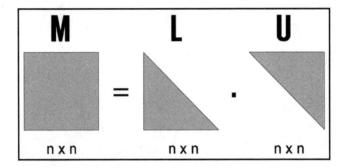

Let's see how it's implemented using `numpy`:

```
from numpy import array
from scipy.linalg import lu

M = np.random.randint(low=0, high=20, size=25).reshape(5,5)
print(M)
# Output
[[18 12 14 15  2]
 [ 4  2 12 18  3]
 [ 9 19  5 16  8]
 [15 19  6 16 11]
 [ 1 19  2 18 17]]

P, L, U = lu(M)
print("P:n {}n".format(P))
print("L:n {}n".format(L))
print("U:n {}".format(U))

# Output
P:
 [[1. 0. 0. 0. 0.]
 [0. 0. 1. 0. 0.]
 [0. 0. 0. 0. 1.]
```

```
[0. 0. 0. 1. 0.]
[0. 1. 0. 0. 0.]]
L:
[[ 1.          0.          0.          0.          0.        ]
 [ 0.05555556  1.          0.          0.          0.        ]
 [ 0.22222222 -0.03636364  1.          0.          0.        ]
 [ 0.83333333  0.49090909 -0.70149254  1.          0.        ]
 [ 0.5         0.70909091 -0.32089552  0.21279832  1.        ]]
U:
[[18.         12.         14.         15.          2.        ]
 [ 0.         18.33333333  1.22222222 17.16666667 16.88888889]
 [ 0.          0.          8.93333333 15.29090909  3.16969697]
 [ 0.          0.          0.          5.79918589  3.26594301]
 [ 0.          0.          0.          0.         -4.65360318]]

P.dot(L).dot(U)
# Output
array([[18., 12., 14., 15.,  2.],
       [ 4.,  2., 12., 18.,  3.],
       [ 9., 19.,  5., 16.,  8.],
       [15., 19.,  6., 16., 11.],
       [ 1., 19.,  2., 18., 17.]])
```

Eigenvalue decomposition

Eigenvalue decomposition is also a decomposition technique that applies to square matrices. When you decompose a square matrix (M) using eigenvalue decomposition, you will get three matrices. One of these matrices (Q) contains eigenvectors in columns, another matrix (L) contains eigenvalues in its diagonal, and the last matrix is a transpose of the matrix of eigenvectors (Q^{-1}).

This can be summed up in the following formula:

$$M=QVQ^{-1}$$

Eigenvalue decomposition will give you the matrices' eigenvalues and eigenvectors.

The following is an illustration of eigenvalue decomposition:

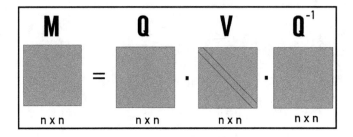

Let's see how it's implemented using numpy:

```
from numpy import array
from numpy.linalg import eig

M = np.random.randint(low=0, high=20, size=25).reshape(5,5)
print(M)
# Output
[[13  9  5  0 12]
 [13  6 11  8 15]
 [16 17 15 12  1]
 [17  8  5  7  5]
 [10  6 18  5 19]]

V, Q = eig(M)
print("Eigenvalues:n {}n".format(V))
print("Eigenvectors:n {}".format(Q))

# Output
Eigenvalues:
 [50.79415691 +0.j          5.76076687+11.52079216j
  5.76076687-11.52079216j -1.15784533 +3.28961651j
 -1.15784533 -3.28961651j]

Eigenvectors:
 [[ 0.34875973+0.j         -0.36831427+0.21725348j -0.36831427-0.21725348j
  -0.40737336-0.19752276j -0.40737336+0.19752276j]
 [ 0.46629571+0.j         -0.08027011-0.03330739j -0.08027011+0.03330739j
   0.58904402+0.j          0.58904402-0.j         ]
 [ 0.50628483+0.j          0.62334823+0.j          0.62334823-0.j
  -0.27738359-0.22063552j -0.27738359+0.22063552j]
 [ 0.33975886+0.j          0.14035596+0.39427693j  0.14035596-0.39427693j
   0.125282  +0.46663129j  0.125282  -0.46663129j]
 [ 0.53774952+0.j         -0.18591079-0.45968785j -0.18591079+0.45968785j
   0.20856874+0.21329768j  0.20856874-0.21329768j]]
```

```
from numpy import diag
from numpy import dot
from numpy.linalg import inv

Q.dot(diag(V)).dot(inv(Q))
# Output
array([[1.30000000e+01-2.88657986e-15j,  9.00000000e+00-2.33146835e-15j,
        5.00000000e+00+2.38697950e-15j,  1.17683641e-14+1.77635684e-15j,
        1.20000000e+01-4.99600361e-16j],
       [1.30000000e+01-4.32986980e-15j,  6.00000000e+00-3.99680289e-15j,
        1.10000000e+01+3.38618023e-15j,  8.00000000e+00+1.72084569e-15j,
        1.50000000e+01-2.77555756e-16j],
       [1.60000000e+01-7.21644966e-15j,  1.70000000e+01-6.66133815e-15j,
        1.50000000e+01+5.71764858e-15j,  1.20000000e+01+2.99760217e-15j,
        1.00000000e+00-6.66133815e-16j],
       [1.70000000e+01-5.27355937e-15j,  8.00000000e+00-3.10862447e-15j,
        5.00000000e+00+4.27435864e-15j,  7.00000000e+00+2.22044605e-15j,
        5.00000000e+00-1.22124533e-15j],
       [1.00000000e+01-3.60822483e-15j,  6.00000000e+00-4.21884749e-15j,
        1.80000000e+01+2.27595720e-15j,  5.00000000e+00+1.55431223e-15j,
        1.90000000e+01+3.88578059e-16j]])
```

QR decomposition

You can decompose a square or rectangular matrix (M) into an orthogonal matrix (Q) and an upper-triangular matrix (R) by applying QR decomposition. This can be expressed in the following formula:

$$M=QR$$

The following is an illustration of QR decomposition:

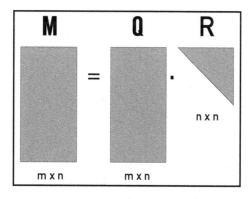

Let's see how it's implemented using numpy:

```
from numpy import array
from numpy.linalg import qr

M = np.random.randint(low=0, high=20, size=20).reshape(4,5)
print(M)
# Output
[[14  6  0 19  3]
 [ 9  6 17  8  8]
 [ 4 13 17  4  4]
 [ 0  0  2  7 11]]

Q, R = qr(M, 'complete')
print("Q:n {}n".format(Q))
print("R:n {}".format(R))

# Output
Q:
 [[-0.81788873  0.28364908 -0.49345895  0.08425845]
 [-0.52578561 -0.01509441  0.83834961 -0.14314877]
 [-0.2336825  -0.95880935 -0.15918031  0.02718015]
 [-0.         -0.          0.16831464  0.98573332]]
R:
 [[-17.11724277 -11.09991852 -12.91095786 -20.68090082  -7.59468109]
 [  0.         -10.85319349 -16.5563638    1.43333978  -3.10504542]
 [  0.           0.          11.88250752  -2.12744187   6.4411599 ]
 [  0.           0.           0.           7.4645743   10.05937231]]
array([[1.40000000e+01, 6.00000000e+00, 1.77635684e-15, 1.90000000e+01,
        3.00000000e+00],
       [9.00000000e+00, 6.00000000e+00, 1.70000000e+01, 8.00000000e+00,
        8.00000000e+00],
       [4.00000000e+00, 1.30000000e+01, 1.70000000e+01, 4.00000000e+00,
        4.00000000e+00],
       [0.00000000e+00, 0.00000000e+00, 2.00000000e+00, 7.00000000e+00,
        1.10000000e+01]])
```

Working with sparse linear systems

You won't always work with dense matrices, and when you need to work with sparse matrices, there are libraries that will help you to optimize sparse matrix operations. Even though these might not have Python APIs, you may need to utilize them by using other programming languages, such as C and C++:

- **Hypre**: Contains preconditioners and solvers to deal with sparse linear systems of equations utilizing parallel implementations.
- **SuperLU**: Deals with large, sparse, nonsymmetric systems of linear equations.
- **UMFPACK**: Solves sparse linear systems.
- **CUSP**: Open source library for sparse linear algebra and graph computations with parallel implementations. By using CUSP, you can access computational resources provided by NVIDIA GPUs.
- **cuSPARSE**: Contains linear algebra subroutines that are used to handle sparse matrices. As with CUSP, you can access computational resources provided by Nvidia GPUs.

Summary

In this chapter, you have explored various low-level libraries that can be paired with NumPy, as well as their configuration. We have deliberately run through the EC2 provision so that you can get familiar with basic Linux command line operations. You have also investigated various compute-intensive, numerical, linear algebra operations that you will use to benchmark different configurations in the next chapter.

In the next chapter, we will create a benchmark python script to be run on each configuration. You will be able to see the performance metrics for different linear algebra operations and different sizes of matrices

Performance Benchmarks

9

In this chapter, you are going to look at the performance statistics of different configurations that we covered in the previous chapter. Of course, the current setup does not provide you the most accurate environment because you have no control over EC2 instances, but it will give you an idea about the setup required in your own environment.

We will be covering the following topics:

- Need for a benchmark
- Performance with BLAS, LAPACK, OpenBLAS, ATLAS and Intel MKL
- Final results

Why do we need a benchmark?

As you advance with your programming skills, you will start to implement more efficient programs. You will search dozens of code repositories to see how others are solving similar problems, and you will find those rare gems that will amaze you.

Throughout this progress of writing better software and implementing systems, you will need ways to measure and track the rate of improvement. You will generally consider your starting point as a baseline and see how the improvements you make will add up to performance metrics.

Once you set the baseline, you will benchmark several different implementations and will have a chance to compare these in terms of the performance metrics you choose. You can choose various metrics and need to decide these in advance.

Performance metrics for these benchmarks will be kept rather simple, and only the time spent metric will be used. You will perform the same operations many times with different configurations and calculate the mean time spent at first. The formula to calculate the mean is:

$$\bar{X} = \frac{1}{n} \sum_{i=1}^{n} X_i$$

This is a good old formula to calculate mean; in our example, the formula is explained as follows:

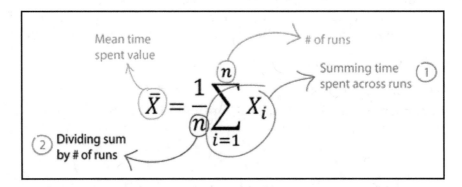

The baseline will be created based on this formula. The first set of calculations will be as follows:

Addition and multiplication of:

- Vector-vector
- Vector-matrix
- Matrix-matrix

You will usually run these calculations a given number of times and calculate the average.

The following code snippet shows you a custom function rather than common timers available in Python. The reason you will use a custom function is that you can extend it later with other statistical functions and have better view of details with proper logging. Function will outputs useful information before the calculation starts and also the results after iterations are finished.

```python
import inspect
import time
from datetime import datetime

def timer(*args, operation, n):
    """
    Returns average time spent
    for given operation and arguments.

    Parameters
    ----------
    *args: list (of numpy.ndarray, numpy.matrixlib.defmatrix.matrix or both)
    one or more numpy vectors or matrices
    operation: function
    numpy or scipy operation to be applied to given arguments
    n: int
    number of iterations to apply given operation
    Returns
    -------
    avg_time_spent: double
    Average time spent to apply given operation
    std_time_spent: double
    Standard deviation of time spent to apply given operation

    Examples
    ---------

    >>> import numpy as np
    >>> vec1 = np.array(np.random.rand(1000))
    >>> vec2 = np.array(np.random.rand(1000))
    >>> args = (vec1, vec2)

    >>> timer(*args, operation=np.dot, n=1000000)
    8.942582607269287e-07
    """

    # Following list will hold the
    # time spent value for each iteration
    time_spent = []

    # Configuration info
```

```
print("""
---------------------------------------------

### {} Operation ###

Arguments Info
--------------
args[0] Dimension: {},
args[0] Shape: {},
args[0] Length: {}
""".format(operation.__name__,
args[0].ndim,
args[0].shape,
len(args[0])))

# If *args length is greater than 1,
# print out the info for second argument
args_len = 0
for i, arg in enumerate(args):
    args_len += 1

if args_len > 1:
    print("""
    args[1] Dimension: {},
    args[1] Shape: {},
    args[1] Length: {}
    """.format(args[1].ndim,
        args[1].shape,
        len(args[1])))

print("""
Operation Info
--------------
Name: {},
Docstring: {}

Iterations Info
---------------
# of iterations: {}""".format(
operation.__name__,
operation.__doc__[:100] +
"... For more info type 'operation?'",
n))

print("""
-> Starting {} of iterations at: {}""".format(n, datetime.now()))

if args_len > 1:
```

```
        for i in range(n):
            start = time.time()
            operation(args[0], args[1])
            time_spent.append(time.time()-start)
    else:
        for i in range(n):
            start = time.time()
            operation(args[0])
            time_spent.append(time.time()-start)

    avg_time_spent = np.sum(time_spent) / n
    print("""
-> Average time spent: {} seconds,
-------------------------------------------
""".format(avg_time_spent))

    return avg_time_spent
```

As you have `Docstring` in this function, you can display it to see function arguments, what it returns, and an example usage:

```
print(timer.__doc__)
```

This will generate the following output:

```
    Returns average time spent
    for given operation and arguments.
    Parameters
    ----------
        *args: list (of numpy.ndarray, numpy.matrixlib.defmatrix.matrix or
both)
            one or more numpy vectors or matrices
        operation: function
            numpy or scipy operation to be applied to given arguments
        n: int
            number of iterations to apply given operation

    Returns
    -------
        avg_time_spent: double
            Average time spent to apply given operation

    Examples
    --------
    >>> import numpy as np

    >>> vec1 = np.array(np.random.rand(1000))
    >>> vec2 = np.array(np.random.rand(1000))
```

```
>>> args = [vec1, vec2]
>>> timer(*args, operation=np.dot, n=1000000)
8.942582607269287e-07
```

Let's start measuring the average time spent for the dot product of 2 vectors. The following code block defines vectors and creates arguments to be fed into a timer function:

```
import numpy as np
vec1 = np.array(np.random.rand(1000))
vec2 = np.array(np.random.rand(1000))
args = [vec1, vec2]
```

You can call the timer function now as follows:

```
timer(*args, operation=np.dot, n=1000000)
------------------------------------------
### dot Operation ###
Arguments Info
--------------
args[0] Dimension: 1,
args[0] Shape: (1000,),
args[0] Length: 1000
args[1] Dimension: 1,
args[1] Shape: (1000,),
args[1] Length: 1000
Operation Info
--------------
Name: dot,
Docstring: dot(a, b, out=None)
Dot product of two arrays. Specifically,
- If both `a` and `b` are 1-D... For more info type 'operation?'
Iterations Info
---------------
# of iterations: 1000000
-> Starting 1000000 of iterations at: 2018-06-09 21:02:51.711211
-> Average time spent: 1.0054986476898194e-06 seconds,
------------------------------------------
1.0054986476898194e-06
```

Our vector-vector product took 1 microsecond on average. Let's see how we can improve this calculation by adding additional metrics. Another metric that you can easily add is standard deviation, which is shown in following formula:

$$sd = \sqrt{\frac{1}{1-n} \sum_{i=1}^{n}(X_i - \bar{X})^2}$$

You are familiar with the formula terms from the previous illustration. Standard deviation simply informs you about the variability of reported metric, which is average time spent in our example.

Extend the timer function by calculating `std_time_spent`, printing its value, and returning the following:

```
avg_time_spent = np.sum(time_spent) / n
std_time_spent = np.std(time_spent)
print("""
-> Average time spent: {} seconds,
-> Std. deviation time spent: {} seconds
""".format(avg_time_spent, std_time_spent))
return avg_time_spent, std_time_spent
```

You can also update the `Docstring` as follows:

```
Returns
-------
avg_time_spent: double
Average time spent to apply given operation
std_time_spent: double
Standard deviation of time spent to apply given operation.
```

You can redefine the time function and run the previous calculation again as follows:

```
timer(*args, operation=np.dot, n=1000000)
```

You will get the following output (just showing the last part for the sake of brevity) with additional information:

```
-> Starting {} of iterations at: {}".format(n, datetime.now())
-> Average time spent: 1.0006928443908692e-06 seconds,
-> Std. deviation time spent: 1.2182541822530471e-06 seconds
(1.0006928443908692e-06, 1.2182541822530471e-06)
```

Great! What other metrics you add? How about adding confidence intervals? That part will be left for you to exercise, but it should be fairly easy for you!

Let's continue with the vector-matrix product:

```
mat1 = np.random.rand(1000,1000)
args = [vec1, mat1]
timer(*args, operation=np.dot, n=1000000)
```

This will give you the following output:

```
Arguments Info
--------------
args[0] Dimension: 1,
args[0] Shape: (1000,),
args[0] Length: 1000
args[1] Dimension: 2,
args[1] Shape: (1000, 1000),
args[1] Length: 1000
Operation Info
--------------
Name: dot,
Docstring: dot(a, b, out=None)
Dot product of two arrays. Specifically,
- If both `a` and `b` are 1-D... For more info type 'operation?'
Iterations Info
---------------
# of iterations: 1000000
-> Starting 1000000 of iterations at: 2018-06-09 19:13:07.013949
-> Average time spent: 0.00020063393139839174 seconds,
-> Std. deviation time spent: 9.579314466482879e-05 seconds
(0.00020063393139839174, 9.579314466482879e-05)
```

Finally, matrix-matrix multiplication is given as follows:

```
mat1 = np.random.rand(100,100)
mat2 = np.random.rand(100,100)
args = [mat1, mat2]
timer(*args, operation=np.dot, n=1000000)
```

This will give you an output similar to previous ones.

Now, we have more or less an idea how challenging these tasks to perform for a computer. List of benchmark functions is complete with dot product added to matrix decomposition you saw in the previous chapter.

What you will do is create a Python script file that includes these calculations and statistics. Then, you will run this file in different configuration that you set up on AWS.

Let's have a look at `linalg_benchmark.py`, which you can find at https://github.com/umitmertcakmak/Mastering_Numerical_Computing_with_NumPy/blob/master/Ch09/linalg_benchmark.py.

The following code block shows you the important part of `linalg_benchmark.py` script, which will be used to test different configurations that you have previously set up on AWS:

```
# Seed for reproducibility
np.random.seed(8053)
dim = 100
n = 10000
v1, v2 = np.array(rand(dim)), np.array(rand(dim))
m1, m2 = rand(dim, dim), rand(dim, dim)
# Vector - Vector Product
args = [v1, v2]
timer(*args, operation=np.dot, n=n)
# Vector - Matrix Product
args = [v1, m1]
timer(*args, operation=np.dot, n=n)
# Matrix - Matrix Product
args = [m1, m2]
timer(*args, operation=np.dot, n=n)
# Singular-value Decomposition
args = [m1]
timer(*args, operation=np.linalg.svd, n=n)
# LU Decomposition
args = [m1]
timer(*args, operation=lu, n=n)
# QR Decomposition
args = [m1]
timer(*args, operation=qr, n=n)
# Cholesky Decomposition
M = np.array([[1, 3, 4],
[2, 13, 15],
[5, 31, 33]])
args = [M]
timer(*args, operation=cholesky, n=n)
# Eigenvalue Decomposition
args = [m1]
```

```
timer(*args, operation=eig, n=n)
print("""
NumPy Configuration:
--------------------
""")
np.__config__.show()
```

There will be two separate runs:

- 1st run with `dim = 100`
- 2nd run with `dim = 500`

Let's have a look at the results.

Preparing for a performance benchmark

For each instance and configuration, navigate to your `Home` directory and create a folder named `py_scripts`:

```
ubuntu@ip-172-31-21-32:~$ cd ~
ubuntu@ip-172-31-21-32:~$ mkdir py_scripts && cd py_scripts
```

Create a file named `linalg_benchmark.py` with the following command and paste the contents:

```
ubuntu@ip-172-31-25-226:~/py_scripts$ vi linalg_benchmark.py
```

After pasting the contents, type :, then type `wq!`, and *Enter* to save and quit:

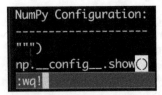

Now you can run this file with the following command:

```
ubuntu@ip-172-31-25-226:~/py_scripts$ python3 linalg_benchmark.py
```

For Anaconda distribution, you will run the script with the following command:

```
ubuntu@ip-172-31-22-134:~/py_scripts$ ~/anaconda3/bin/python linalg_benchmark.py
```

Performance with BLAS and LAPACK

Here, you will run the `linalg_benchmark.py` script with BLAS and LAPACK. Connect to the **t2.micro** instance where you have this configuration, and run the script as shown in the previous section.

The following are the results of the run with `dim = 100`:

Operation	Mean	Std. Deviation
V-V Product	0.00000122	0.00000071
V-M Product	0.00000872	0.00000147
M-M Product	0.00074976	0.00001754
SV Decomp.	0.00644510	0.00009101
LU Decomp.	0.00042435	0.00001801
QR Decomp.	0.00134417	0.00003373
Cholesky D.	0.00001229	0.00000306
Eigval Dec.	0.01133923	0.00014564

The following are the results of the run with `dim = 500`:

Operation	Mean	Std. Deviation
V-V Product	0.00000169	0.00000104
V-M Product	0.00018053	0.00001345
M-M Product	0.09042594	0.00078627
SV Decomp.	1.72078687	2.11683465
LU Decomp.	0.36958391	0.05764444
QR Decomp.	1.64355660	0.26008436
Cholesky D.	0.00012395	0.00203646
Eigval Dec.	11.03387896	1.19246878

Performance with OpenBLAS

Here, you will run `linalg_benchmark.py` script with OpenBLAS. Connect to **t2.micro** instance where you have this configuration, and run the script as shown in the previous section.

The following are the results of the run with `dim = 100`:

Operation	Mean	Std. Deviation
V-V Product	0.00000115	0.00000059
V-M Product	0.00000333	0.00000135
M-M Product	0.00009168	0.00000847
SV Decomp.	0.00507356	0.00005898
LU Decomp.	0.00016124	0.00001763
QR Decomp.	0.00065833	0.00001702
Cholesky D.	0.00001366	0.00000374
Eigval Dec.	0.03457905	0.00043139

The following are the results of the run with `dim = 500`:

Operation	Mean	Std. Deviation
V-V Product	0.00000124	0.00000078
V-M Product	0.00006752	0.00000487
M-M Product	0.00752822	0.00009364
SV Decomp.	0.13901888	0.00128025
LU Decomp.	0.00575469	0.00009780
QR Decomp.	0.02157722	0.00024035
Cholesky D.	0.00001288	0.00000212
Eigval Dec.	3.94406696	3.75736472

Performance with ATLAS

Here, you will run the `linalg_benchmark.py` script with ATLAS. Connect to **t2.micro** instance where you have this configuration, run the script as shown in previous section.

The following are the results of the run with `dim = 100`:

Operation	Mean	Std. Deviation
V-V Product	0.00000118	0.00000078
V-M Product	0.00000537	0.00001443
M-M Product	0.00029508	0.00011157
SV Decomp.	0.00475364	0.00025615
LU Decomp.	0.00015830	0.00000738
QR Decomp.	0.00093086	0.00004695
Cholesky D.	0.00001311	0.00000290
Eigval Dec.	0.01048062	0.00028431

The following are the results of the run with `dim = 500`:

Operation	Mean	Std. Deviation
V-V Product	0.00000168	0.00000054
V-M Product	0.00013248	0.00001036
M-M Product	0.02474427	0.00063530
SV Decomp.	0.22419701	0.00352764
LU Decomp.	0.00561713	0.00013463
QR Decomp.	0.05162554	0.00122877
Cholesky D.	0.00001262	0.00000260
Eigval Dec.	3.18629725	2.77181242

Performance with Intel MKL

Here, you will run `linalg_benchmark.py` script with Intel MKL. Connect to **t2.micro** instance where you have this configuration, and run the script as shown in the previous section.

The following are the results of the run with `dim = 100`:

Operation	Mean	Std. Deviation
V-V Product	0.00000432	0.00031263
V-M Product	0.00000357	0.00005485
M-M Product	0.00007010	0.00035516
SV Decomp.	0.00241478	0.00065733
LU Decomp.	0.00015441	0.00008672
QR Decomp.	0.00055125	0.00030522
Cholesky D.	0.00001264	0.00003074
Eigval Dec.	0.00746131	0.00012120

The following are the results of the run with `dim = 500`:

Operation	Mean	Std. Deviation
V-V Product	0.00000140	0.00001808
V-M Product	0.00006262	0.00000957
M-M Product	0.00670626	0.00009224
SV Decomp.	0.09701678	0.00102559
LU Decomp.	0.00496843	0.00010792
QR Decomp.	0.01590121	0.00027027
Cholesky D.	0.00001278	0.00000220
Eigval Dec.	0.22408283	0.00155203

Results

Of course, the **t2.micro** instance is fairly weak and you should know more about how Amazon provides this computing power for EC2 instances. You can read more about them at https://aws.amazon.com/ec2/instance-types/.

If you use more powerful machines with a higher number of cores, the performance difference will be more visible between different configurations.

When it comes to results, it's no surprise that the default installation of BLAS and LAPACK gave us the baseline performance and optimized versions, such as OpenBLAS, ATLAS, and Intel MKL, gave better performance.

As you have noted, you haven't changed a single line of code in your Python script and by just linking your NumPy library against different accelerators, you had huge performance gains.

If you will dig deeper into these low-level libraries to understand what are the specific routines and functions are provided, you will have a better idea of which parts of your program will benefit from these implementations.

Of course, there are many other details that you might not understand at first. It might be the case that the functions you're using are not using the low-level library or doesn't parallelize the operations. There might be some cases where multithreading will or will not help. Knowledge and experience ultimately depend on your experiments, and you will be more proficient with various applications as you will learn from your own experience.

Many researchers published the design and results of their experiments. A quick Google search will give you a bunch of resources to read and understand how these libraries perform with different hardware and software configurations.

Summary

In this chapter, you explored the performance of different configurations when you perform compute-intensive linear algebra operations.

Benchmarking is a serious business, and you at least have the basic skills now to run benchmarks. The material you have studied in this chapter is nowhere near complete, but it gave you an idea where to start, and you can definitely improve on many things.

One thing you can look at is how performance metrics behave when you increase the size of vectors and matrices gradually. Ideally, you'll need more powerful hardware, but **t2.micro** instances are free in most cases or very cheap to provision.

As you will need to handle more compute-intensive workloads, it's important to understand what your options are, and which one will give you the best performance. You can run these kinds of simple experiment to at least have an idea about the performance, and it will help you a lot and save time and money.

If you have come this far, congratulations! We believe going through all the chapters and studying the materials advanced your skills when it comes to Python scientific stack.

We hope you enjoyed reading this book, and we would like to thank you for your time.

Other Books You May Enjoy

If you enjoyed this book, you may be interested in these other books by Packt:

SciPy Recipes

L. Felipe Martins, Ruben Oliva Ramos, V Kishore Ayyadevara

ISBN: 9781788291460

- Get a solid foundation in scientific computing using Python
- Master common tasks related to SciPy and associated libraries such as NumPy, pandas, and matplotlib
- Perform mathematical operations such as linear algebra and work with the statistical and probability functions in SciPy
- Master advanced computing such as Discrete Fourier Transform and K-means with the SciPy Stack
- Implement data wrangling tasks efficiently using pandas
- Visualize your data through various graphs and charts using matplotlib

Python Data Analysis - Second Edition
Armando Fandango

ISBN: 9781787127487

- Install open source Python modules such NumPy, SciPy, Pandas, Statsmodels, scikit-learn, theano, keras, and tensorflow on various platforms
- Prepare and clean your data, and use it for exploratory analysis
- Manipulate your data with Pandas
- Retrieve and store your data from RDBMS, NoSQL, and distributed file systems such as HDFS and HDF5
- Visualize your data with open source libraries such as matplotlib, bokeh, and plotly
- Learn about various machine learning methods such as supervised, unsupervised, probabilistic, and Bayesian
- Understand signal processing and time series data analysis
- Get to grips with graph processing and social network analysis

Leave a review - let other readers know what you think

Please share your thoughts on this book with others by leaving a review on the site that you bought it from. If you purchased the book from Amazon, please leave us an honest review on this book's Amazon page. This is vital so that other potential readers can see and use your unbiased opinion to make purchasing decisions, we can understand what our customers think about our products, and our authors can see your feedback on the title that they have worked with Packt to create. It will only take a few minutes of your time, but is valuable to other potential customers, our authors, and Packt. Thank you!

Index

www.ingramcontent.com/pod-product-compliance
Lightning Source LLC
LaVergne TN
LVHW081521050326
832903LV00025B/1581